✱ Minimal plastic was used in the printing of this book.

How I Kicked the Plastic Habit and How *You* Can Too

BETH TERRY

FOREWORD BY JACK JOHNSON

Skyhorse Publishing

Skyhorse Publishing books may be purchased in bulk at special discounts for sales promotion, corporate gifts, fund-raising, or educational purposes. Special editions can also be created to specifications. For details, contact the Special Sales Department, Skyhorse Publishing, 307 West 36th Street, 11th Floor, New York, NY 10018 or info@skyhorsepublishing.com.

www.skyhorsepublishing.com

10 9 8 7 6 5 4 3

Library of Congress Cataloging-in-Publication Data is available on file.
Print ISBN: 978-1-63220-665-7
Ebook ISBN: 978-1-63450-035-7

Jacket design by Jane Sheppard
Interior design by Brian Peterson

Printed on "ECO-G" Recycled Woodfree Paper, made of 100 percent recycled pulp (post-consumer and pre-consumer mixed).

Printed in China

For Michael Stoler, my hero.

In memory of Betty Terry, who didn't know the meaning of the word procrastination.

Publisher's Note on Making a Plastic-Free Book

What do we mean by minimal plastic?

When we decided to publish a book called *Plastic-Free*, we knew we wanted to make the book itself plastic-free. As it turns out, that's something of a challenge—with plastic coating on the cover and jacket, polyester or nylon thread in the binding, and plastic glue throughout, most books are full of plastic! So we've stripped things down. The cover and jacket are uncoated, the thread is made of cotton, and the spine is exposed. Our printer even managed to find a plastic-free glue to use. With all that in mind, we assure you that if the book's not 100 percent free of plastic, it's as close as can be!

Author's Note on the Updated Edition

Please download a free reader's guide for your book group at
www.MyPlasticFreeLife.com/book/

In the original 2012 edition of *Plastic-Free*, I wrote, "While the general principles and strategies of plastic-free living are timeless, the specifics may change. Companies go out of business and new ones emerge. Researchers make discoveries. Websites move. After reading this book, please visit MyPlasticFreeLife.com for in-depth reviews of new plastic-free products and solutions, posts on the effects of plastics on human health and the environment, up-to-date resource pages, and stories and suggestions from the community on ways to reduce plastic in everyday life."

What I didn't anticipate back then was just how frequently web addresses change. So, in this new edition, I have chosen to omit most URLs. But don't worry: Google is our friend. Every reference in this book can be found online simply by typing the name into a search engine, which is what most of us do nowadays anyway.

I also didn't realize at the time just how American-centered the book was. Since then, I've been contacted by people all over the world wondering what they can do to reduce plastic where they live. The fact is, there are now bloggers and activists across the globe writing about living plastic-free—people in Europe, Asia, Australia, New Zealand, and probably more that I haven't yet discovered. So I've added a whole new chapter to the book to introduce you to a few of them and hopefully inspire you to get started reducing plastic wherever you live.

And I've kept learning and researching new and better plastic-free alternatives. There are whole new sections in this edition (plastic-free sunscreen, for instance) and changes and updates to previous sections as I've revised my opinions or gathered additional information. And there is updated information on a few of the characters introduced in the first edition: Jean Hill, for example, with her anti–bottled water campaign. Please enjoy this new, updated edition of *Plastic-Free*.

Beth Terry
Facebook: beth.terry1
Twitter: @PlasticfreeBeth

Contents

Foreword

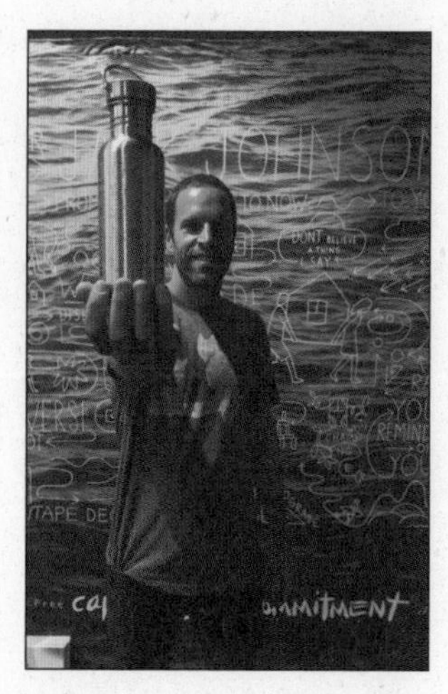

I am excited to see a second edition of *Plastic-Free* by Beth Terry. Beth's book, with its ingenuity, hands-on examples, and humor, will serve as an inspiration and educational resource to readers. The pages that follow provide a toolkit for actionable behavior change at home, work, and in the wider world. Beth's work mirrors the personal and professional commitments I have made to reduce my waste footprint, spread the plastic-free message, and support policy change.

My initial concerns about our addiction to plastic were inspired by a lifelong love of the ocean. Since the printing of the first edition of *Plastic-Free*, Beth has

visited our home on O'ahu, Hawai'i to share her work with school groups and the wider community. She also saw firsthand how plastic pollution in the form of marine debris is affecting our beaches when attending a beach cleanup hosted by Kōkua Hawai'i Foundation.

My wife and I try to live life with less plastic. In *Plastic-Free*, Beth details a number of practices that we actually follow at home such as using glass food storage containers, investing in durable or upcycled goods, and always carrying our reusable water bottles, straws, and utensils. We understand that our choices today will impact our children's future.

As Beth shares, when measured collectively, these personal behavior changes make a big difference. On tour, my band and crew make every effort to reduce waste and single-use disposable plastics, and we have tried to positively extend such practices to others. We have eliminated plastic water bottles backstage and provide water refill stations so people can fill reusable bottles at the shows. Fans are asked to take action with our "Capture Your Commitment" campaign and share a plastic-free pledge on social media. We created the All At Once social action network and have partnered with hundreds of non-profits around the world who raise awareness about alternatives to disposable plastic and encourage positive change. We support these non-profit partners through the Johnson Ohana Charitable Foundation, including Tangaroa Blue Foundation and their Australian Marine Debris Initiative, and Algalita Marine Research & Education and their youth action summit focused on solutions to plastic pollution. We support innovative school projects such as Community Environmental Council's "Rethink the Drink" program which installs hydrations stations in Santa Barbara, and Cafeteria Culture's efforts to get rid of Styrofoam trays in New York City. We have seen an incredible rise in awareness and a growing momentum toward solutions and alternatives. Readers will find sound advice on how to nurture similar behaviors and successful community campaigns in *Plastic-Free*.

In 2003, my wife Kim and I started the Kōkua Hawai'i Foundation to support environmental education in Hawai'i's schools and communities. Kōkua Hawai'i Foundation programs include 'ĀINA In Schools, 3Rs School Recycling, KHF Field Trip Grants, KHF Mini-Grants, and Plastic Free Hawai'i. The Plastic Free Hawai'i program specifically provides resources, tools, and trainings to educate schools, business partners, and community members about the environmental and health benefits of going plastic-free to minimize single-use plastics. Our program engages the public via educational

outreach booths at various community and partner events such as local farmers' markets and youth surf competitions as well as in-school presentations. Resources for this type of work are offered in greater detail in Beth's book, such as packing a waste-free lunch, upcycling an old t-shirt into a tote bag, supporting businesses that use less packaging, and much more.

Again, we have seen how small actions add up. In the past three years, Plastic Free Hawai'i volunteers have personally encouraged over 11,500 Hawai'i students, adult residents, and visitors to make a plastic-free commitment, such as to choose reusable bags and containers or give up bottled water. Nearly 5,000 participants have attended Plastic Free Hawai'i beach cleanups and collected over 18,000 pounds of trash and marine debris from O'ahu beaches. In partnership with Method, some of this recovered marine debris is being recycled into packaging for the world's first soap bottle made from ocean plastic. We have supported legislation initiatives making Hawai'i the first state to ban single-use plastic bags at checkout in all counties, and recently distributed 15,000 reusable tote bags at our outreach events across the island of O'ahu to help with the transition. Plastic Free Hawai'i has additionally hosted youth and educator summits, film screenings, and presentations from experts in the field, such as Charles Moore and Beth herself.

Read, enjoy, and become part of a broader movement. I hope this book inspires you as it has inspired us.

Aloha, Jack

Portrait of Beth Terry by collage artist Tess Felix, made entirely from plastic litter collected from Stinson Beach, CA. Her website is tessfelixartist.com.

Introduction: Waking Up to Plastic

Hi. My name is Beth Terry, and I am an addict.

Before June of 2007, I lived the plastic lifestyle. It's no great surprise. Most of us do; it's pretty standard in the United States. It's a lifestyle of consumption, enabled by convenience. We can get pretty much anything we want, or think we want, pretty much whenever we want it. We don't have to think about the costs, beyond what is printed on the package or on our receipts. We don't have to think about the consequences. And all this can exist, in part, because of plastic: products are inexpensive because they are not designed to last, they are packaged so that they can wait indefinitely on store shelves for our sudden desires, and they can be thrown away and forgotten.

Back then, "moderation" was a foreign word to me. I ate junk food to excess, obsessing daily about my next chocolate fix and making sure I had a plastic-wrapped stash on hand to satisfy the next craving. I lived on coffee all day (in disposable cups with plastic lids), wine all night, and very little sleep in between. Take-out food in Styrofoam clamshells and frozen macaroni and cheese dinners were my best friends.

As for recreation, I collected CDs and DVDs competitively, stockpiled scrapbook materials, and, since I was training for a marathon, accumulated heaps of synthetic technical clothing and training tools I thought were absolutely essential. In between, I barely had time to do my accounting job, much less go to bed. Fortunately, I had pills to help me sleep when staying up all night playing computer games or blogging or ordering stuff online became too overwhelming. Anything else I might need was a short walk away at the late night convenience store, or even just "1-Click" on Amazon.com, which was sure to bring me a satisfyingly huge box, in which the precious object of my desires would be nestled, protectively swaddled in multiple layers of wrapping and surrounded by plastic air pillows. At least in material terms (for I was a material girl, living in a material world), life seemed good, and, to paraphrase the slogan of the American Chemistry Council—the voice of the plastics industry—plastic made it possible. I generated bins full of plastic trash, but what did I care? I didn't see where it went. It just disappeared to that magic place called "away."

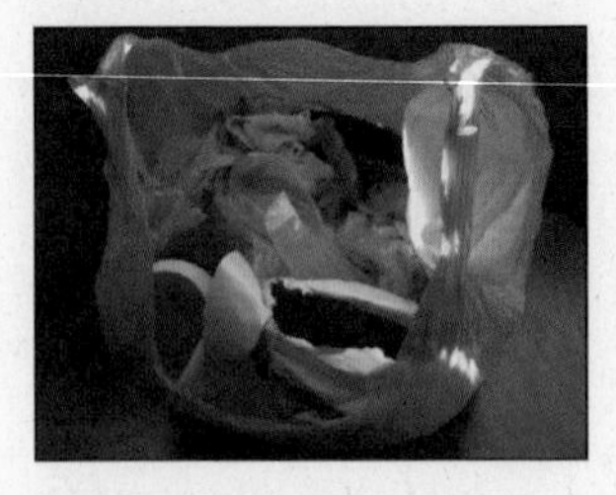

"How could a convenience addict like me go from generating huge trash bags of the stuff each week to just one small grocery bag full in all of 2011?"

Fast forward to today. For the past five years I've been living with almost no new plastic. How could a convenience addict like me go from generating huge trash bags of the stuff each week to just one small grocery bag full in all of 2011? And how did giving up plastic help me recognize and tame some of my more urgent addictions in the process?

It started with surgery.

In the summer of 2007, I was stuck at home for several weeks recuperating from an operation. No worries: I was prepared. My kitchen was stocked with plenty of bottled water and juices, frozen microwaveable meals, and energy bars, conveniently delivered via Safeway.com. And I was plugged in, ready to stave off boredom with the television and iPod and, of course, computer. What's more, just in case Netflix, iTunes, and YouTube weren't enough, I was surrounded by shelves of books I'd collected and never read. I could have been laid up in our apartment for months and never needed to leave.

Except that something had disrupted the mail service that summer, and my Netflix DVDs were not arriving. Despite the fact that I already had shelves full of unwatched DVDs and unread books, I called the post office day after day, becoming more and more anxious and irritated. "You don't understand!" I insisted frantically on Day Four of this ordeal. "I've just had a ***hysterectomy*** and I have no new movies to watch! I need my fix!" I was forced to seek stimulation elsewhere, such as on public radio, and one evening I heard a story about a man in New York named Colin Beavan, who had spent a year trying to live in a way that would have as little negative effect on the environment as possible. He was writing a book about it, called *No Impact Man*, and maintained a blog of the same title. Interested, I began browsing through his site, following links from it. I soon came to an article by journalist Susan Casey, entitled "Our Oceans are Turning Into Plastic . . . Are We?"—in, of all places, the online version of the magazine *Men's Health*—and the shocking photo that would change everything.

The picture showed the decomposed carcass of a Laysan albatross, an ungainly-looking sea bird that nests on Midway Island, which is halfway between California and Japan surrounded by thousands of miles of Pacific Ocean. The flesh of this particular bird—a chick!—had fallen away to reveal a rib cage filled with plastic bottle caps, disposable cigarette lighters, even a toothbrush—small pieces of plastic that had no business out there in the middle of nowhere. Pieces of plastic like those I myself used and tossed away every day.

Frozen in my desk chair, I stared at the awful image. For several seconds, I literally could not breathe.

And then, I forced myself to read the entire article. Tragically, this chick was not unique. Thousands of albatross mothers mistake tiny plastic pieces for food floating on the surface of the ocean. They swallow them up from the waters of the North Pacific Gyre, an area between the United States and Japan that is increasingly becoming known as the "Great Pacific Garbage Patch" because of all the plastic waste collecting there, and fly back to Midway to feed this "food" to their chicks. Except that plastic is not food. Huge numbers of baby albatrosses die of starvation each year, their bellies full of the dross of human civilization—the stuff that you and I throw away casually every day. And while the body of the bird will finally disintegrate and return to the earth, the plastic that killed it will linger on in the environment, never biodegrading, available once again to be eaten by future birds, so that the deadly cycle would continue.

As I thought of how I used and tossed away pieces of plastic just like these every day, I felt as if, like Coleridge's ancient mariner, I had helped kill this albatross.

Laysan albatross chick carcass filled with plastic trash

I knew my life had to change. And that, like the mariner, I had to tell people about the albatross.

What was it about that particular story that stopped me dead in my over-consuming, mindlessly addicted tracks? I'd seen sad animal photos before. Fairly recently I had thrown money at the Sierra Club and Greenpeace. And I'd watched *An Inconvenient Truth* like a good San

Francisco Bay Area liberal, misting up over polar bears stranded on melting ice. But nothing affected me as personally as the photo of that dead albatross.

Perhaps in that moment, I would have been just as stricken by any baby animal in danger. I had, after all, just lost my uterus, and the realization that I'd never have kids of my own, despite having thought I didn't want them in the first place, was starting to sink in. The utter permanence of it. I was grieving. And trying to distract myself from grieving. (Where were my freaking Netflix movies?!?) Realizing the emptiness that my life had become, I sat and stared at the screen for half an hour, letting my heart break.

And then I got mad. Why wasn't anyone doing anything to fix this problem? Clean it up? Where was our government? Why had I never heard of this "plastic garbage patch" from any of the tree-hugging organizations that sent me mailings several times a week? Not that I had time to read their depressing missives anyway. But as I read more of this article, I learned that it's not just birds eating our waste, but fish and other marine life that consume bits of plastic—many of those pieces coated with toxic chemicals—and pass the pollution up the food chain to us. We're pumping this stuff into the environment so fast that it's coming back to us on our dinner plates.

I learned that chemicals from plastics could leach into our food and drinks and that children's plastic toys, dishes, and sippy cups could contain hormone-disrupting additives. Not only was the well-being of other animals at stake, but also our own health and that of our families and friends. Making the connection, I realized that this plague of plastic chemicals was harming the most vulnerable members of our planet—children and animals—and that was both unacceptable and unfair.

But what could I do? I knew that while I couldn't personally go out and clean up the Gyre, I could start with myself. And then I started to remember . . . There was a time when I really did think that what I did was important. In kindergarten, I wrote a play that was performed by my class. It was called *Ploshin Ploshin* (my phonetic spelling of "Pollution, Pollution"), and it had songs and everything. The play was about three characters, Betty, Beth, and Billy, who suddenly notice that there is garbage everywhere and decide to pick it all up. To my five-year-old mind, pollution meant litter. I grew up with anti-litter television commercials like Woodsy Owl's "Give a Hoot; Don't Pollute" campaign from the U.S. Forest Service or the crying Indian commercial from the Keep America Beautiful (KAB) campaign. I didn't know then that KAB was actually a cynical move by the bottled beverage industry to blame the litter problem on consumers. I just knew that litter was bad, and it was up to me to help clean it up.

As a child, I believed that my actions mattered. Raised in the Mormon Church, I was taught to speak up for what was right, despite what other people might think. And although I left the Church as a teenager, that belief—that my actions were important—led me to a short-lived job working for the environmental group Clean Water Action after graduating from college. Canvassing door-to-door to protect the environment, having the courage to speak to complete strangers, and maintaining the conviction that we could make a difference, I thought I was unstoppable. I took it upon myself to organize a letter-writing campaign against Tampax when the company introduced tampons with plastic applicators in the late 1980s, and I even got the guys in the office to participate.

And then, somehow, I lost those early convictions. The practicalities of making a living took over. My priorities changed. By the time I saw the photo of that dead albatross chick, I had become the kind of person who would choose double plastic bags at the grocery store on purpose and even gloat a little about her post-tree-hugger status. I was someone who would buy a new plastic bottle of water every time she went to the gym and toss it in the trash because there were no recycling bins handy. And when I heard that a certain low-priced yuppie grocery store chain was opening up in our neighborhood, I sang for joy each morning in the shower for all the plastic-packaged salads I was going to be able to buy on my way to work. Sadly, the store opened after I had my plastic-awakening, and I've not had the same enthusiasm for it since.

The day I saw that photo, I committed to looking at my own plastic consumption and plastic waste and figuring out what changes I could make. How much plastic was I using (I had no idea), and how much could I actually give up? Off on medical leave with plenty of time on my hands, I dived into my new project with my usual obsessive intensity. Vowing to collect all of my plastic trash each week, I set up a blog (originally called *Fake Plastic Fish*, because if we don't stop filling up our oceans with plastic, they could be the only kind of fish we have left) to keep myself on track.

Further, I vowed to buy as little new plastic as possible, to instead find plastic-free alternatives and to share these discoveries on my blog. To make sure I had eyes watching and holding me accountable, I spammed each and every one of my friends and family with my blog posts, sometimes several a day.

A few weeks later, I'd recovered physically and returned to my job as an accountant for a Bay Area home care agency, but by this time, ***Fake Plastic Fish*** had taken on a life of its own. I'd linked up with other green bloggers, like "No Impact Man" Colin Beavan, scheduled visits with recycling centers to find out where our plastic recycling goes, become involved with environmental organizations, like the Bay Area environmental group Green Sangha, that are actively working to educate the public to rethink their relationship to plastic, and even planted the seeds for a future consumer action campaign. By day, I continued my accounting job. But nights and weekends were consumed by plastic!

In the years since my plastic awakening, I've gone from personally generating almost four pounds of plastic waste per month to a little over two pounds per year (the average American generates between 88 and 120 pounds per year, and that's only what they throw away at home!), and I am continuing the downward trend. While I've learned many facts about plastic—how it's made, which types leach toxic chemicals into our food, why plastic recycling is actually "downcycling," and the many ways in which it's both helpful and harmful—the biggest lessons have been personal rather than factual.

As I used up the plastic products I already had, I'd research ways to replace them without buying new plastic. Some products were easy to replace (solid soap bars packaged in paper rather than liquid soap in plastic bottles), others were more difficult (plastic-free recycled toilet paper), and still others were impossible to replace. These impossibilities, while initially irritating, ultimately taught me a lot about my way of life. The fact that there were, in the end, no plastic-free replacements for the frozen convenience foods and energy bars I had come to rely on forced me to take a deeper look at my assumptions about eating and to rethink what I was putting into my body.

And as I examined how I was spending (wasting) resources and money, I realized that the way I was spending my time was no longer sustainable either. I had replaced several addictions with one great big one. How could I argue for giving up plastic-wrapped addictions like convenience foods or Hershey Bars and plastic-molded products like brand new computers and iPhones and flat screen TVs when I myself was staying up all night blogging about plastic and running myself into the ground just as surely as I had been doing before? It just didn't make sense. Sleep, I have learned, is one of the best things we can do for the planet, if only because we're not consuming much while we do it. And oh yeah, it's good for our bodies and minds, too.

No one's perfect, least of all me. This journey is ongoing. I didn't vow to go without plastic for a year only to return to my old ways of being. I'm in it for the long haul. And I still experience some of the frustrations and challenges I did back in 2007. I didn't write this book to tell anyone what to do, but as an invitation to join me in this journey of personal and ecological discovery. Sure, in all honesty I do want to inspire you and your friends and family to use less plastic. But more than that, to learn what it is about plastic that makes it the symbol of what Captain Charles Moore (discoverer of the Great Pacific Garbage Patch) calls the "crisis of our civilization," and figure out ways to get out from under the thumb of plastic addiction.

I'll share what I've learned about plastic so far: what it is, how it's manufactured, in what products it's found (some may surprise you!), and what health and environmental issues are associated with the different types of plastic. I'll interview a few experts and even some explorers. But the main focus of this book is solutions rather than problems.

The movement to reduce overconsumption and pollution stresses the "4 Rs": refuse, reduce, reuse, and recycle. Re***fuse***—the verb, not to be confused with the noun ***ref***use, or garbage, though that is, in effect, what we are refusing—means that we don't ***have*** to use plastic. We can opt for other sorts of products, ones without plastic components or plastic packaging. Throughout this book, I'll suggest alternatives to plastic products, many of which follow simple formulas and can be whipped up at home. Or we can choose to change our lifestyles so as to eliminate whole classes of products entirely. (In many cases, this is a lot easier than it sounds.)

Reduce simply means using less. It's nearly impossible to eliminate plastic from our lives entirely. And, unlike me, you may not even want to try. That's OK. This isn't a competition, and I don't hold myself up as an ideal. Again, the goal is to be aware and responsible. You might find yourself saying, "I know this is plastic and bad for the environment (and for my health), but I really need it or want it right now." You think about the consequences, weigh your options, and make a decision. Other times, you may say to yourself, "It might be nice to have this, but it just isn't worth the pollution and health risks. I can do without it." Perhaps you will set a goal—to reduce your plastic consumption by a certain percentage or keep it under a certain amount per month. Or you might define what constitutes an emergency situation for you and save plastic for those times. Whatever you do, it will be *your* choice.

Reuse means to reject the culture of single-use—the idea that it's normal, natural, even required, to throw away perfectly good, durable items after one use. Dieters say of certain foods, "Seconds on the lips, forever on the hips"; something similar may be said about many plastic items, such as eating utensils: seconds in your hand, forever in the land(fill). On the other hand, reuse can sometimes be a problem with plastic—we'll see why.

The fourth R, recycle, is an important last resort, but it can also be problematic when it comes to plastic, as we'll see. I'll describe my visit to a local recycling center and explain why recycling is not the best solution to our plastic problem. This is why it's always last on the list, the least preferable alternative, a way to try to clean up the damage, while we try to prevent it from being done in the first place.

But beyond these 4 Rs, there are others. One simple one is replace, that is, substitute nonplastic products for those that use plastic, which makes it a lot easier to refuse. We'll take a look, for instance, at bioplastics—and I'll explain why those, while a step in the right direction, are not the entire solution either.

We can also remember. There was a time when plastic wasn't omnipresent. If you are above a certain age, your parents may recall such a time, or you may. *Not everything* was made of plastic—and yet we survived and turned out all right. Possibly better. Yes, times have changed, and some plastics have made our lives safer, but we can still find useful guidance from the way we used to live. My husband Michael's mother, a diligent reader of my blog, is always telling him, with a slightly triumphant laugh, that when she sees some of my suggestions, she thinks, "But I've *always* done that." She doesn't need to stop using plastic because she never started. Again, by stopping to think and stepping off the track that carries us along and tells us that a certain lifestyle is the only "modern," correct one, we can get back to one that makes more sense.

Another R is repair. Instead of automatically chucking out broken appliances, ripped clothing, and obsolete technology, we can find ways to refurbish what we already have. In previous generations, products were made to last and could easily be repaired when they broke. Services existed with seamstresses, mechanics, and technicians trained to restore our property to working order. Nowadays, it's often cheaper to replace our dead gadgets than to repair them, but the price of new products doesn't take into account the environmental costs to dispose of the old and manufacture the new. With a little ingenuity, we can take matters into our own hands and find ways to lengthen the lifespan of the stuff we already have.

A fun R is report. For the past several years, I've been reporting my findings on my blog, and now via this book, sharing information about the problems and the solutions. You can do the same. There are many forums for letting others know what we have learned and discovered about plastic-free living—including my own blog at www.MyPlasticFreeLife.com—a growing community of people looking at their personal plastic consumption and working towards reducing it. It's great to know we're not alone.

And the next R, the next step, is rally. Some people argue that changes in our plastic consumption habits aren't really going to change the world. The problem is just too big, the ocean too polluted, and the amount of greenhouse gases emitted too enormous. How could one person using a canvas bag instead of a plastic one really have any effect? But large groups of people united in a cause can influence companies, industries, and political institutions. I'll discuss several instances in which organizing in a cause made a difference. Our personal changes can lead to whole movements.

But perhaps the most important R precedes and permeates all the others. And that is realize. Living plastic-free is inextricably entangled with awareness, consciousness, mindfulness. In a world of plastic, we don't have to think about where the products we use come from or where they're going. Our lives are unexamined. Like my cats, we can loll around all day, being cute, secure in the knowledge that food and shelter will be provided for us from sources whose nature we don't really have to worry about. This is great for cats, but not for humans. Humans make decisions, humans take responsibility—another R.

These principles will show up throughout my story, a chronicle of my own fits and starts figuring out how to live with less plastic. Like a teenager, I began my journey very self-righteous and thought I knew everything. I can be bullheaded—just ask my husband, Michael. By running up against my own assumptions, weaknesses, and addictions time after time, I've learned a few things that might be helpful to you. At the very least, I hope my story is entertaining.

This book is also a practical guide to reducing our plastic consumption. There are so many books and movies out these days alerting people to the environmental problems we face but few offering real, practical solutions. I don't have all the answers. But I do have a lot of them! And in addition to sharing ways I have found to live with less plastic, this book offers strategies for dealing with the roadblocks that inevitably present themselves any time we try to make important changes in our lives. About a year or so into my less-plastic challenge, a journalist asked if I ever got embarrassed bringing my own bags and containers shopping with me. I said no, but the truth is that swimming against the current can be hard, as it challenges our usual ways of functioning in the world. It's nice to have a little help.

And finally, this book is a starting point for activism. In order to stem the tide of plastic pollution, we need big actions from companies and governments. But in order to get there, we each have to start with ourselves. Systemic change comes from the collective will of individuals: you and me, wanting to make a difference, starting with ourselves and learning how to go further. I hope that by the end of this book, you'll be ready to take on the world, or at least your local city council and grocery store.

If I've learned nothing else in these past several years, it's that plastic is not the enemy. It's a symptom and symbol of an attitude of obliviousness to what is really going on in our world, in our lives, in our bodies, and of a way of life that is no longer sustainable. If we can examine and change the fundamental ways we view ourselves and our relationship to the rest of the world, we can solve the plastic problem from the ground up. Or at least we can give it a damned good try.

Chapter 1: Plastic Is Everywhere

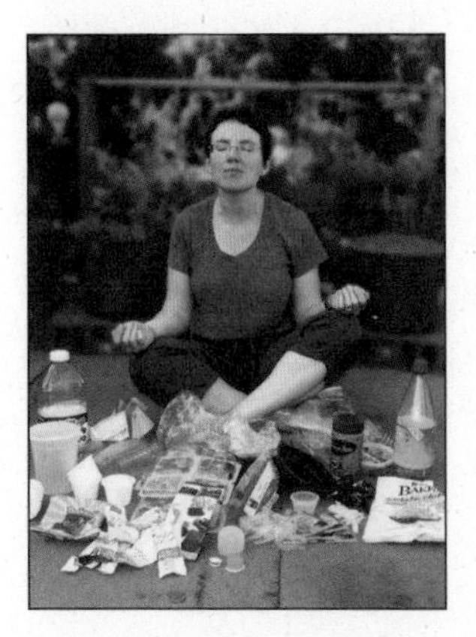

The day after my plastic epiphany, I went to work collecting my plastic waste. I set up a paper grocery bag under the kitchen table and began tossing every used up plastic container, wrapper, cup, spoon, tray, or plastic package into it, along with anything broken, torn, or just plain unusable. I assured Michael right off the bat that this was my project and that he was not required to participate with me. I think he was a little wary at first. He knew how my interests could quickly become obsessions and wondered if I'd keep my promise about not nagging him to join. It was either that, or he assumed this was just one more of Beth's wacky ideas that would be over within a few months. But he didn't comment, just watched with amusement from the sidelines as I struggled to figure out what the hell I was doing. The bag filled up much more quickly than I expected, and with the mounting pile of plastic came questions: Do I count plastic I purchased this week but haven't used up yet? Do I count plastic I already had before I started this project? What about plastic that I receive as gifts?

I could see I would need some guidelines to navigate my new plastic-free life. I wanted to try to live without buying any new plastic, but at that point, I wasn't interested in getting rid of the plastic I already had. And I knew I couldn't possibly give up all plastic all at once. I had to decide which plastics were essential and what I could let go of.

My Rules

1. I will collect all my plastic trash (both recyclable and non-) each week, photograph and tally it up on a spreadsheet, and post the list to my blog. I will *only* include plastic waste that I generate or contribute to. If Michael brings home food packaged in plastic and I consume some of it, the packaging will go into my tally. But anything he buys for his exclusive use (like cottage cheese in plastic tubs!) will not count.

2. I will continue to use up the plastic-packaged products I already have but will try to avoid buying any new plastic. Using up what I have will buy me the time I need to research alternatives.

3. If there is a plastic item I really need, I can borrow or rent or acquire it second hand. I just can't buy it new. And I can't cheat by accepting new plastic gifts from friends.

4. I will avoid storing or eating food in plastic containers because of chemicals that might leach from them.

First Month's Plastic Tally

You might assume I would have been afraid to look at my first few weeks' plastic collections—afraid of how much plastic trash I had generated. But I wasn't. In fact, I was excited to finally have some idea of my personal plastic footprint. I intended to examine each item as a scientist would: objectively, without guilt or blame. Feeling guilt would only get in the way of learning. I had already decided I was going to start limiting my plastic consumption. What good would it do to beat myself up over what was already done, and done in ignorance? The first week I dumped out that paper grocery sack full of plastic trash onto my back deck and started sorting through it was almost thrilling. And as an accountant, I loved creating lists and charts and spreadsheets. This was going to be fun.

In all, I collected roughly 3.5 pounds of plastic trash in the first month. That's more than the amount of plastic waste I generated in all of 2011! And it's not even the complete list. Other items not included in the tally were a huge bag full of plastic grocery bags that I intended to reuse and disposable water bottles I had tossed away at the gym. And there were items I didn't include because I didn't realize they were plastic at the time, such as plastic wine "corks" or cardboard cups, cartons, and containers coated with plastic (like many of my frozen dinner trays).

So, what would I do with all this stuff? Well, the first thing I figured I should do is find out what could be recycled and put that stuff in the recycle bin. I had been carting all of the plastic that couldn't be recycled in my city of Oakland, California, into San

Francisco to dump into my friends' recycle bin. But now, I wanted to make sure I was doing the right thing, that the items going into the bin were actually being recycled. I needed to understand what the differences were among all the various types of plastic. And just what was this stuff called plastic in the first place?

What Is Plastic?

Here's a little chemistry lesson. Despite my nonscientific background, I find the story of plastic fascinating, and I hope you will too.

Plastic is a polymer. But not all polymers are plastics, and not all polymers are problematic. In fact, natural polymers are actually the building blocks of life! Giant molecules made up of repeating chains of smaller molecules called monomers and composed primarily of carbon and hydrogen ("organic compounds"), polymers are found throughout nature. The starches, proteins, and DNA in our bodies are polymers. Shell, horn, fingernail, and hair are made of polymers. Cellulose, the primary component in the cell walls of green plants, is a polymer, as is natural rubber. Resins such as shellac, secreted by the lac bug and often used as a furniture finish, and amber, a fossilized tree resin, are polymers. Because natural polymers have evolved over millions of years alongside the planet's living systems, such polymers biodegrade and return to the earth.

> Collecting my trash was enlightening, as you'll see throughout this book. If you think you'd like to take the challenge and get an idea of how much plastic waste you generate, check out the Show Your Plastic Challenge section at the end of this chapter.

For over a century, scientists and inventors have understood the usefulness of polymers and have developed methods for manipulating the qualities of these polymers to create plastics. In 1855, the inventor Alexander Parkes devised the very first plastic, called Parkesine (later called celluloid), from natural cellulose. He did it by treating the cellulose with nitric acid and a solvent, then dissolving the result in alcohol and hardening it into a transparent and elastic material that could be molded when heated. Other plastic products developed from

In the first four weeks, I collected:

- 1 ripped grocery bag
- 3 sparkling water bottles (#1 PET)
- 1 spring water bottle (#1 PET)
- 16 energy drink mix packets
- 1 Brita faucet filter cartridge
- 1 take-out soup container and lid (#4 LDPE)
- 2 clamshell containers for take-out food (#6 PS)
- 2 drinking straws
- 2 plastic spoons
- 1 small salad dressing to-go cup & lid
- 1 fortune cookie wrapper
- 1 soy sauce packet
- 1 toothpick with plastic on the end
- 1 juice bottle (#1 PET)
- 3 pudding cups (#5 PP)
- 2 hummus containers (#5 PP)
- 9 granola/energy bar wrappers
- 14 mini candy wrappers
- 2 chip bags
- 4 cheese wrappers
- 1 grated parmesan cheese container and lid
- 1 English muffin bag
- 1 pita bread bag
- 1 raisin bag
- 1 plastic tea box wrapper
- 1 cracker box wrapper
- 1 Costco spaghetti sauce multi-pack wrapper
- 2 spaghetti wrappers
- 10 plastic films from frozen dinners
- 4 frozen dinner trays (#1 PET)
- 1 frozen vegetable bag
- 1 ice cream sandwich wrapper
- 4 soy milk spouts and caps
- 1 yeast packet
- 1 foam insert from vitamin bottle cap
- 1 lid and pump from a can of shave gel
- 1 lid and pump from a can of air freshener
- 1 plastic dental brush pick box
- 1 dental brush pick
- 1 mouthwash bottle and wrapper (#3 PVC)
- 2 soap bar wrappers
- 1 liquid soap bottle and pump
- 28 single-use eye drop vials (#4 LDPE)
- 27 expired single-use eye drop vials (#4 LDPE)
- 1 eye drop box wrapper
- 1 expired saline solution bottle (#2 HDPE)
- 1 feminine hygiene product bag
- 18 Invisalign dental aligners
- 15 Invisalign aligner blister packs
- 1 toothpaste tube and cap

- 1 tiny plastic palette and spatula from a container of facial bleach
- 2 toilet paper package wrappers
- 1 paper towel package wrapper
- 1 facial tissue box wrapper
- 1 plastic mount from a Venus shaver cartridge container
- 1 wrapper from a package of sponges
- 1 used scrubber sponge
- 1 broken plastic dish wand & sponge
- 1 blister pack from a new candy thermometer
- 1 broken plastic food storage container
- 1 piece of gift wrapping ribbon
- iPod accessory packaging
- 2 electronics wrapper bags
- 1 strip of Scotch tape
- 1 blister pack from a glue stick (and the glue stick itself)
- 1 dried up ballpoint pen
- 1 CD wrapper
- 1 plastic card holder from a bouquet of flowers
- About 9 feet of used packing tape
- 1 FedEx padded pak
- 1 Ziploc-type bag
- 10 air pillows packing material
- 6 unidentifiable clear plastic wrappers

Beth Terry with collected plastic from first six months of 2007.

cellulose were cellophane (used for clear plastic tape, bags, and wrappers), and smokeless gunpowder. By manipulating natural polymers, inventors were able to create a whole range of plastic products, and since all of these products were based on natural polymers, they were all biodegradable.

And then, in 1909, Alexander Baekeland invented the very first purely synthetic polymer, Bakelite. Bakelite was a whole new polymer that had never existed in nature. It was made from phenol and formaldehyde, two chemicals that Baekeland derived from coal. And since it had not evolved for millions of years like natural polymers, there was no organism that could break it down. Thus came into existence the non-biodegradable plastics we use today.

As I mentioned, plastics are derived from carbon and hydrogen—hydrocarbons—and the majority of hydrocarbons are found in fossil sources, such as crude oil, coal, and natural gas. Of the hundreds of petroleum-derived hydrocarbons, the most useful for making plastic are methane, ethylene, propylene, butylene, and benzene. And these chemicals are not just used to make plastics but also soaps, detergents, fertilizers, pesticides, and even cosmetics! Tens of thousands of chemicals are made from combinations of these simple molecules, and many of these chemicals have never been tested for human safety. But I'll get into that later. For now, I just want to focus on plastics, those long repeating chains of hydrocarbons and other molecules—materials that humans have created—that have never before existed in nature, and that therefore cannot be broken down by any known microorganism back into elements found in nature. For a comprehensive explanation of the chemistry and history of plastic, told in an engaging and accessible style, pick up a copy of Susan Freinkel's book *Plastic: A Toxic Love Story* (Houghton Miffliin Harcourt, 2011).

A Walk Through My House

Nowadays, so many products are made from synthetic polymers that it's hard to imagine a world without them. And would we even want to? Looking up from the computer screen on Day 1, I suddenly realized that plastic is everywhere. The computer itself, of course, and its speakers, mouse, keyboard, and printer. But also the pens on my desk, the stack of software CDs on the shelf, and the covers of the 3-ring binders lined up next to them. My backpack on the floor next to my desk and the desk chair in which I had just been sitting. Moving from the office to the living room, I saw lamps made of plastic, my TV and DVD player, and all the music and movie discs I'd accumulated. In the hallway were my umbrella, raincoat, athletic shoes, and boots. In the kitchen, I noticed the plastic knobs on the stove and toaster oven. The base of my blender and the pitcher on my food processor. My dishwasher, microwave, and refrigerator were completely lined with plastic. My cupboards were full of plastic food containers and Teflon-coated pans. And of course the refrigerator and pantry were full of foods packaged in plastic: containers, wrappers, bags, bottles, jars, as well as lids and caps. I noticed the vacuum cleaner, the sponge mop and bucket, and the broom with its plastic handle. Under the sink, I spied various cleaning products in plastic bottles, as well as synthetic sponges and brushes. In the bathroom, I noticed my comb, brush, and hair dryer. In the cabinet: toothpaste, toothbrush, dental floss, medicines in plastic bottles, shampoo and conditioner, liquid soap, lotions and creams, and cosmetics. We didn't have a plastic shower curtain, but that's only because we had glass shower doors. In the bedroom, I noticed drawers full of synthetic clothing, a clock radio, pillows stuffed with polyester. I remembered that up in the attic were all my camping supplies: tent, tarp, backpack, sleeping bag. And throughout the apartment from wall to wall—synthetic carpet. We were living in a plastic world. How could I possibly live completely plastic-free? I wasn't sure I wanted to. But for the purposes of my challenge, I was going to have to find ways to avoid buying any of these things new.

Problems with Plastic

The plastic items in my house, as well as the wrappers and bottles and containers and bags I had collected in the grocery bag under the kitchen table, all looked and felt so different from each other. It occurred to me that some were probably more harmful than others. I wanted to learn if there were any safer plastics. But as I researched the many issues related to plastic, I learned that while some are worse than others, there are troubling issues associated with all of them.

Let me just say up front that the information in the next few pages is a bummer. Some of the environmental problems we've created in the name of convenience will be hard to solve on a global level—hard, but not impossible. Keep in mind that this is a book of solutions, and I firmly believe that if we each start with ourselves, we can create a wave of change to reverse the precarious position in which we now find our world. What's more, living with less plastic can make us personally healthier and even happier. So let's first tackle the beast head on and then proceed to the fun stuff.

1. Plastic is Made from Fossil Sources, Such as Petroleum and Natural Gas

The raw materials for plastic—petroleum and natural gas—must be extracted from the ground, using machinery and techniques that can devastate the surrounding terrain. Oil can spill and contaminate vast areas of land and water. On April 20, 2010, BP's Deepwater Horizon oil rig exploded, causing the biggest oil spill in U.S. history. Millions of gallons of oil gushed up from the earth over several months, destroying the livelihoods of people and the ecosystem throughout an entire region. When I first learned about the disaster, I was numb for days. The tragedy was too big to comprehend. People talked about boycotting BP and going across the street to buy their gas from BP's competitors, and they blamed BP's executives. But I wonder how many people stopped to think about all the ways in which we support the oil industry in the first place. We mainly consume oil as a transportation fuel, as diesel, kerosene for jet planes, and gasoline for cars. But even when we reduce our travel miles, which is essential for cutting our dependence on oil, our lives are still full of petroleum because so many of the synthetic chemicals we use on a daily basis are made from the stuff—including plastics, ubiquitous plastics.

The plastics industry points out that much of the plastic in the United States is made not from oil but from "clean," domestic natural gas. But drilling for natural gas is anything but "clean." According to the National Institutes of Health, both oil and gas production can emit hazardous air pollutants, including carcinogens like benzene, toluene, xylenes and other hazardous chemicals, including nitrogen oxides, volatile organic compounds, carbon monoxide, sulfur dioxide, and particulate matter.[1] What's more, drilling for gas allows methane to escape into the atmosphere, where it is a cause of global warming up to thirty-three times worse than carbon dioxide. Recently, a new method of extracting natural gas has become popular. Called hydraulic fracturing, or "fracking," it involves breaking up formations of gas-containing rock by injecting huge amounts of water, mixed with hundreds of proprietary chemicals of unknown biological effect, into the ground, where they can contaminate drinking water supplies. And unfortunately, a provision of the Energy Policy Act of 2005 exempts most fracking fluids from regulation under the Safe Drinking Water Act, making it nearly impossible for the U.S. Environmental Protection Agency to protect our water supplies from these chemicals.[2]

So whether we are fighting wars to secure our claim to nonrenewable petroleum; polluting our air, water, and earth through emissions from the petrochemical industry; or raising the temperature of our planet by releasing the carbon that was once bound up in fossil deposits, we must take a hard look at how our dependence on these resources threatens life on this planet. Of course, we need to reduce the amount of fuel we use for driving and powering our homes. That's a given. But reducing our demand for plastics is one more step toward creating a world beyond fossil fuels. The good news is we can do it.

2. Plastic Contains Toxic Chemicals

All plastic products contain a plethora of additives, which can include fillers, pigments, stabilizers, flame retardants, antistatic additives, plasticizers, blowing agents, lubricants, antibacterials, fungicides, fragrances, and more. These chemicals are added to the basic polymer to affect its strength, texture, flexibility, color, resistance to microbes, and other characteristics. These chemicals can leach out of the plastic into our food and beverages, and some can off-gas into the air we breathe and even rub off from the objects we handle. Some chemicals added to plastic or used in the production of plastics are "endocrine disruptors," which mimic the natural hormones our bodies produce, increasing the risks of early puberty in females, reduced sperm counts, altered functions of reproductive organs, altered sex-specific behaviors, obesity, diabetes, and even some breast, ovarian, testicular, and prostate cancers.[3] Unlike most toxic chemicals, endocrine disruptors may actually have an ***increased*** effect in very small doses,[4] effects that can begin in the womb. (To understand how chemicals can have greater effects at smaller doses, see "The Low Dose Response" below.[5]) And the effects of all the toxic chemicals in plastics are multiplied in children because of their smaller, but growing, bodies.[6] Here are some of the toxic chemicals we already know are in certain plastics:

THE LOW DOSE RESPONSE

Why is there so much debate about whether endocrine disrupting chemicals in plastics, like BPA or phthalates, are harmful to humans in the amounts at which we're exposed to them? And why is it so difficult to pass regulations in the United States to protect us from these chemicals? One reason is that regulators are accustomed to following the age-old adage that "the dose makes the poison." They study chemicals at high doses to see what the toxic effects are and decrease the dose until they don't see any more toxic effects on the cells. They then assume that below that threshold, a chemical is safe. However, when it comes to endocrine disruptors, that threshold is just the beginning.

To understand this phenomenon, I spoke with Dr. Laura Vandenberg, then researching endocrine disruptors at Tufts University, who explained it in a way a non-scientist like me can comprehend. At doses lower than the toxic threshold, endocrine disruptors mimic the hormones in our bodies. And hormones—chemicals such as insulin, thyroxin, estrogen, and testosterone—regulate vital functions like body growth, response to stress, sexual development and behavior, production and utilization of insulin, rate of metabolism, intelligence and behavior, and the ability to reproduce. The interactions occur through a number of mechanisms, the easiest of which to imagine is the lock and key.

The cells in our body contain hormone receptors, molecules that bind to a particular hormone. You can think of the receptors being the locks and the hormones being the keys to specific locks. Chemicals that mimic hormones in the body will bind to those receptors and create effects in the body similar to hormones. Those effects can increase the likelihood of diseases like obesity, cardio vascular disease, reproductive disorders, behavioral disorders, autism, diabetes, and various cancers.

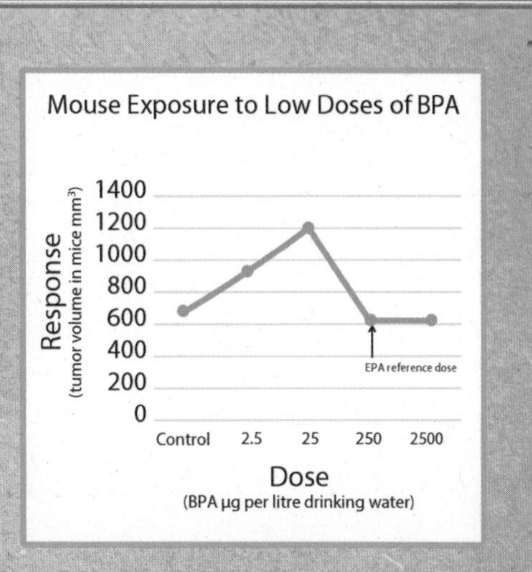

A 2011 experiment shows tumors in mice at BPA doses lower than the equivalent EPA safe threshold for humans. Data source: Jenkins, S., et al. Environmental Health Perspectives 119, 1604 – 1609, Figure 1 and Figure 2C (2011).

But the important thing to note is that the body only has so many hormone receptors. So once a high enough dose of a chemical is reached, the receptors will be saturated, meaning that no further hormone response will be seen, no matter how much higher the dose of the chemical. At even higher doses, it becomes toxic for the receptor to keep responding, so it shuts down, and we actually see fewer effects than we did at lower doses.

What about the effects of these chemicals at even smaller doses? None of us is exposed to one chemical at a time but a whole Long Island iced tea of chemicals on a daily basis. So even if one particular chemical is below the threshold for observed effects, the combination of chemicals can increase their effects.

The bottom line is that chemicals like BPA, phthalates, parabens, PCB, DDT, and other endocrine-disruptors can be dangerous for us no matter what the dose. At high levels, they are acutely toxic. At low levels, they disrupt the endocrine system. And at even lower levels, they can still combine with other chemicals in the environment to disrupt the endocrine system. As citizens, we need to understand how these chemicals work so that we are not mislead when the chemical industry claims that a particular product is safe because the amount of a particular chemical in it is too low to make a difference. That kind of reasoning is just too simplistic.

Bisphenol-A (BPA) is a component of hard, polycarbonate plastic historically used to make reusable water bottles, baby bottles, sippy cups—although due to public pressure, manufacturers have been phasing out BPA for those products—5-gallon water jugs, food storage containers, children's dishes, utensils, and tub toys. It's used in the epoxy lining of metal food and beverage cans (including beer and soda cans, as well as cans of children's food products), the linings of metal lids and screw caps, the coatings of thermal paper cash register receipts, and even some dental sealants. In animal studies, BPA has been shown to be a reproductive, developmental, and systemic toxicant, and it is a known endocrine-disruptor.[7] To date, a lot of BPA concern has been for its affect on children's growing bodies, but in 2011, researchers from California Pacific Medical Center found that the chemical interferes with breast cancer drugs.[8] Several U.S. states (including my state of California) have banned BPA in certain types of plastic products. In 2009, Canada banned BPA from baby bottles, and in 2011, the European Union followed suit. In 2012, the U.S. FDA amended its regulations to no longer provide for the use of BPA-based polycarbonate resins in baby bottles and sippy cups, and in 2013, it amended its regulations to no longer provide for the use of BPA-based epoxy resins as coatings in packaging for infant formula. But these steps were taken only after the industry had voluntarily banned these uses of the chemical in response to public pressure. As of this writing, neither the FDA nor EPA has restricted any other uses of BPA and it continues to proliferate.[9]

BPA-Free Alternatives are not necessarily safer than BPA. Because of public outcry against BPA, many companies have developed BPA-free alternatives, proprietary formulas which have not been sufficiently tested for safety. Nowadays, the labels of many children's products and water bottles scream BPA-FREE in bold type. But several studies comparing products with BPA to their BPA-free counterparts found that many of the BPA-free products, including water bottles and baby bottles, produced more "Estrogenic Activity" (the ability to mimic the hormone estrogen in the body) than those containing BPA. By exposing the plastics in these products to human breast cancer cells, scientists determined that many of the BPA-free substitutes produced the same or greater hormone-disrupting effects as BPA.[10] It's not enough for us to know what's *not* in a plastic product; as citizens, we should have a right to know what *is* in it.

Phthalates are a class of chemicals added to polyvinyl chloride (PVC, a.k.a. vinyl) to make it soft. PVC has been used to make a whole host of products, including cling wrap, some squeeze bottles—although as with BPA, manufacturers have been switching to phthalate-free alternatives for food wraps and food containers due to public pressure—lunch boxes, binder covers, children's toys, purses, rainwear, shower curtains, hoses, soft tubing, medical blood and IV bags, vinyl flooring, window blinds, house siding, the plastic coating on electrical wires, window frames, and other soft plastic products. In addition to PVC, phthalates have been found to leach from polyethylene terephthalate (PET) bottles, and while the PET industry insists that no phthalates are used in the manufacture of PET (despite its name), phthalates are somehow finding their way into even that "safer" plastic.[11] Certain phthalates like DEHP have been linked to birth defects, reproductive disorders, asthma, autism, and various cancers. And because phthalates are not chemically bound to the plastics in which they are added, they can be inhaled via phthalate-contaminated dust and even absorbed into the skin through touch.[12] While the Consumer Product Safety Improvement Act (CPSIA) of 2008 limits the use of several types of phthalates in children's toys and child care articles (DEHP, DBP, and BBP, as well as DINP, DIDP, and DnOP, pending further study), it does not prohibit their use in all consumer products.[13] Currently, many consumer products contain these chemicals, and manufacturers are not required to disclose their presence.

Lead and Cadmium are heavy metals sometimes used to stabilize PVC plastic, and both have been found to leach out of children's plastic toys, older mini-blinds, vinyl flooring, and other products.[14] Lead can damage the nervous system, kidneys, blood, and brain and is reasonably anticipated to be a human carcinogen.[15] Cadmium is a known human carcinogen and can damage the kidneys irreparably.[16] The CPSIA limits the use of lead, cadmium, and other heavy metals in children's products; however, other consumer products, as well older PVC children's products, may still contain those additives.[17]

Styrene is a component of polystyrene plastic and expanded polystyrene foam (e.g., Styrofoam) which is often found in disposable food containers, disposable cups (insulated foam cups, as well as the smooth red and blue plastic Solo cups so ubiquitous at parties), and even hard plastic utensils. In 2011, the U.S. Department of Health and Human Services' National Toxicology Program added styrene to its list of chemicals "reasonably anticipated to be a human carcinogen." (The listing was upheld by the National Academy of Sciences in 2014.) Workers in plastic production plants are particularly susceptible to styrene exposure, but styrene has also been found to leach in smaller amounts from polystyrene food packaging.[18]

Antimony is a catalyst used in manufacturing polyethylene terephthalate plastic (PET), the kind of plastic in the clear disposable plastic bottles used for water, soda, juices, other beverages; and other disposable food containers. Recent studies in the E.U. have found concentrations of antimony in fruit juices bottled in PET plastic well above the legal standards for drinking water.[19] Antimony is a toxic metal that plays no biological role in the human body.[20] One recent study suggests it may even be an endocrine disruptor[21]. But whether or not the long-term effects of antimony are harmful in the amounts present in water and juice is the subject of ongoing research.

Animal Fat is sometimes used as a "slip agent" to lubricate plastics and keep them from sticking to metal surfaces during production. Plastic bags may contain slip agents to prevent them from sticking together. For this reason, vegetarians may want to avoid plastic products as much as possible.[22]

Antibacterial Chemicals like Triclosan and Microban (which may contain Triclosan) are often added to household plastics such as cutting boards and food containers, as well as shower curtains, sponges, toothbrushes, toys, clothing, and other consumer products, to prevent the buildup and growth of bacteria on surfaces. Some of these products are labeled as such, and some are not. The market for antibacterial additives for plastics is growing, according to a recent marketing study by the Helmut Kaiser Consulting Firm.[23] According to the Environmental Working Group, Triclosan is linked to liver and inhalation toxicity, and low levels of the chemical may disrupt thyroid function.[24] But besides the direct health effects from leaching of such chemicals, the move to add antibacterial chemicals to all of our household products has the undesired effect of creating ever more potent strains of super bugs that are increasingly resistant to the chemicals we use to kill them. What's more, antibacterial chemicals kill not only the bad bugs but the good bacteria that help us fight them off. And preventing our kids from exposure to germs early in life compromises their developing immune systems, which need a certain amount of exposure to microbes to learn how to respond to them.

Toxic Flame Retardants Halogenated chemicals (those which contain bromine or chlorine bound to carbon) like polybrominated diphenyl ethers (PBDEs) and chlorinated Tris—the latter of which was actually banned from children's pajamas in the

1970s—are flame retardants often added to electronics, plastics, wiring, insulation, and anything containing polyurethane foam (a kind of plastic), including upholstered furniture, nursing pillows, strollers, baby carriers, other foam baby products, and foam carpet padding.[25] At face value, the use of flame retardants might seem like a reasonable safety precaution, given the high flammability of petroleum-based plastics, until you look into the science of these chemicals, which have been shown to cause reproductive, thyroid, endocrine, developmental and neurological disorders, decreased fertility, birth defects, learning disorders, and hyperactivity.[26] These chemicals can leak out of products and contaminate house dust. Until recently, many furniture manufacturers used toxic flame retardants in order to pass California's flammability standard TB-117, whether all of their furniture was sold in California or not. But in December 2013, a new California standard was passed, one which measurably increases fire safety without the use of toxic chemicals. Furniture complying with the new regulations will bear a tag that reads, "TB 117-2013"; however, that tag does not necessarily mean that the furniture is free of flame retardants since the new regulation does not actually ban the use of those chemicals. It's certainly important to keep ourselves safe from fires, but studies have shown that California's previous standard did not measurably increase fire safety in the first place.[27] I'll discuss some naturally flame retardant alternatives in chapter 9.

Perfluorinated Compounds (PFCs) are polymers used to keep foods from sticking. Teflon coating, for example, is a PFC, as are the coatings on many paper wrappers for burgers, sandwiches, microwave popcorn, and even butter. A recent study concluded that PFCs used to coat paper wrappers can migrate into food and enter the human blood stream.[28] Long-chain PFCs bioaccumulate in wildlife and humans, and they are persistent in the environment. They are likely human carcinogens, and they produce reproductive, developmental, and systemic effects in animal laboratory tests. Exposure to PFCs before birth has been linked to lower birth rate in both animals and humans. And the EPA anticipates that given the long half-life of these chemicals in humans (years) and the fact that we are exposed to them throughout our lives, beginning in the womb, continued exposure could increase our body burdens to harmful levels.[29]

Other Chemicals In addition to those listed above, there are over a thousand other chemicals added to plastic or used in the production of plastic. How do we know which of them are safe? Unfortunately, in the United States, we as consumers don't know.

Doesn't our government keep us safe? Unlike many European countries, the United States does not follow the Precautionary Principle,[30] which dictates that when an activity raises the threat of harm to human health and the environment, we should take precautionary measures to protect ourselves even before all the scientific data is in. What's more, the Precautionary Principle specifies that companies should bear the burden of proving their products are safe ***before*** putting them on the market. In other words, we and our children should not have to be the guinea pigs for corporations' latest and greatest chemical offerings. Sadly, that is just what happens in the United States.

But even worse, the United States does not even require manufacturers to disclose to the public what chemicals they add to their plastics in the first place. The intention is to help companies protect their "trade secrets." So it becomes ***doubly*** impossible to determine whether a particular plastic item is safe, since even if we knew whether a particular additive was harmful or not, we wouldn't necessarily know if it was in that particular kind of plastic or not.

So, for instance, because of chemicals like BPA, phthalates, heavy metals, styrene, and antimony, experts warn us to avoid #3, #6, #7 (polycarbonate) and to a lesser extent #1 plastics. (Please refer to the chart at the end of this chapter for explanation of the numbers.) But I was shocked to learn that back in 2008, a group of scientists in Alberta, Canada, accidentally discovered chemicals leaching from polypropylene (#5, one of the "safe" plastics). These scientists were not studying plastic itself, but they happened to be using plastic test tubes, which had always been assumed not to leach. Finding that their results had been contaminated, they studied the test tubes and found chemicals like antibacterials and animal-based slip agents leaching from the plastic. They'd had no idea these chemicals were in the plastic, much less that they could leach out.[31]

So are there any safe plastics? Given what we know (which, as we've seen, is not a whole lot), some ***seem*** to be safer than others:

- #2 high density polyethylene (HDPE), #4 low density polyethylene (LDPE), and #5 polypropylene (PP) are generally considered to be safer. (See SPI Resin Identification Code Chart at the end of this chapter.)
- #1 polyethylene terephthalate (PET) is not recommended to be reused, as it is not easy to clean and readily breaks down under stress, which may result in chemicals leaching out.
- #3 polyvinyl chloride (PVC), #6 polystyrene (PS), and #7 polycarbonate (PC) should be avoided.
- Steer clear of the newer BPA-free plastics that have not yet been proven to be safe.
- Avoid perfluorinated compounds contained in non-stick cookware such as Teflon, as well as in food wrappers and dental tape.
- Don't subject plastic food containers to the high heat of the microwave, dishwasher, or the hot trunk of a car.

But keep in mind that as long as companies are not required to disclose their proprietary formulas, and as long as all additives are not subject to rigorous testing, we can't be sure that any plastics are safe. Until we have laws mandating testing of chemicals ***before*** they enter the market, and disclosure of additives by manufacturers, we would be wise to avoid all plastic products and packaging whenever possible.

For more information about avoiding the worst chemicals in plastics and what you can do to advocate for updated toxics legislation, check out the following websites:

- **Safer Chemicals, Healthy Families** is a coalition of public health organizations, health care providers, environmental organizations, mom bloggers, parent groups, state and community organizations, and businesses pushing for smart federal policies to protect us from toxic chemicals.

- **Safer States** is part of the Safer Chemicals, Healthy Families group, which pushes for updated toxics legislation on the state level.
- **Environmental Working Group** researches chemicals in many different types of products, publishes reports which publicize the toxics we're regularly exposed to as well as consumer guides to safer products, and maintains a huge searchable index of chemicals.
- **Center for Health, Environment, & Justice's PVC Campaign** includes a wealth of information about the dangers of PVC plastic, as well as an annual *Guide to PVC-free School Supplies*; the This Vinyl School interactive website revealing all the places PVC can be hiding in your child's school; a PVC-free University Activist Toolkit; and much more.
- **HealthyStuff.org** is a project of the Michigan-based Ecology Center, which researches and reports on toxic chemicals in everyday products and publishes consumer action guides.
- **Center for Environmental Health** works to hold companies accountable for the toxic chemicals they produce, through outreach, education, and when necessary, litigation.
- **Healthy Child, Healthy World** posts regular articles on practical ways to avoid toxic chemicals in your home and provides tips, checklists, and guides for toxic-free living.
- **Green Science Policy Institute** educates consumers on the science and dangers of halogenated flame retardants and lists ways you can avoid toxic flame retardants in household and children's products while keeping your home safe from fires.
- **The Endocrine Disruption Exchange, Inc.** (TEDX) Founded by the late Dr. Theo Colborn, TEDX is the only organization that focuses primarily on the human health and environmental problems caused by low-dose and/or ambient exposure to endocrine disruptors, including many additives in plastics.
- **Breast Cancer Fund** works to expose and eliminate environmental causes of breast cancer, including some of the chemicals in plastics.

- **Health Care Without Harm** is an international coalition working to get toxic chemicals—including PVC and phthalates—out of hospitals.
- **Environmental Health News** (www.environmentalhealthnews.org) A fantastic source of scientific information on plastics in the environment and their effects on human health.

Books

- ***Big Green Purse: Use Your Spending Power to Create a Cleaner, Greener World,*** by Diane MacEachern. Avery, 2008.
- ***Smart Mama's Green Guide: Simple Steps to Reduce Your Child's Toxic Chemical Exposure,*** by Jennifer Taggart. Center Street, 2009.
- ***Slow Death by Rubber Duck: How the Toxic Chemistry of Everyday Life Affects Our Health,*** by Rick Smith and Bruce Lourie. Knopf Canada, 2009.
- ***Toxin Toxout: Getting Harmful Chemicals Out of Our Bodies and Our World,*** by Bruce Lourie and Rick Smith. St. Martin's Press; Reprint edition, May 6, 2014.

3. Plastic Manufacturing Plants Harm Workers and Pollute Communities

In addition to the toxic chemicals that can harm consumers of plastic products, toxic emissions from plastics manufacturing plants threaten the health and well-being of workers and residents of the communities in which those plants operate.

For example, during production, PVC plants can release dioxins, suspected carcinogens that bioaccumulate in humans and wildlife and are associated with reproductive and immune system disorders. In fact, residents of Louisiana, which is home to half the PVC production facilities in the United States, have been shown to have much higher concentrations of dioxins in their blood than the average U.S. citizen. And sadly, these chemical plants tend to be located in the poorest towns. For example, in the predominantly African American town of Mossville, Louisiana, surrounded by fourteen petrochemical plants, where residents suffer from diabetes, liver cancer, kidney problems, and where the incidence of endometriosis is so high that disproportionate numbers of young women have had to undergo total hysterectomies, blood tests have revealed three times the national average of dioxin in their bodies.[32]

Gulf communities of Southeast Texas like Houston and Corpus Christi, the hotspots of the petrochemical industry, are regularly subjected to disproportionate levels of benzene (a carcinogenic building block of plastics like PET, linked to leukemia and other blood disorders) and other hazardous chemicals used in the production of plastics and petroleum-based products, above safe standards.[33]

So even if the plastic products we use ourselves were proven not to leach chemicals into our own food and beverages, the manufacture of those products still contributes to the pollution of the planet and its poorest citizens. Plastic pollution, it turns out, is also a social justice issue. Is a little convenience worth that price?

4. Plastic is Not Biodegradable

The plastic that we throw away never actually goes away. As I mentioned before, synthetic polymers have only been around for about a century. They are giant molecules that have never before existed in nature, so there's no microbe as yet that can break them down. For this reason, they will last in the environment virtually forever. In the best case scenario, plastics might be recycled (which, you'll learn, is actually downcycling) or sent away, by our overburdened waste management system, to a landfill or an incinerator, for "disposal." But where is "away"?

A landfill is basically a big hole in the ground, lined with a clay and/or plastic liner. Modern landfill liners are meant to last many years, but according to the EPA and other environmental quality experts, even the best liners will eventually wear out, allowing their toxic contents to leach out and possibly contaminate groundwater supplies.[34] Modern landfills are meant to be "dry tombs," preventing even organic matter like food or yard waste from biodegrading. (In fact, excavators at Fresh Kills Landfill on Staten Island have recovered perfectly preserved 30-year-old newspapers and hot dogs.)[35] Biodegradation is undesirable in the anaerobic conditions of a landfill because the process creates methane, a more potent greenhouse gas than carbon dioxide. And while many landfills have begun to capture and convert methane gas to energy, no systems can prevent some leakage into the atmosphere. But, you might argue, plastics don't biodegrade to begin with, so they won't create methane. True. But consider the waste of energy and resources to extract fossil fuels, ship them to refineries, convert them to disposable plastics, only to be used once and sit entombed in the landfill forever, where they may eventually leach their toxic chemicals into the groundwater. It's not a pretty picture.

But burning plastics in an incinerator is not a great option either, as incinerators can produce toxic emissions that poison our air, release greenhouse gases like carbon dioxide and nitrous oxide that contribute to global climate change, and generate toxic ash that will end up in the landfill anyway. In fact, a recent study by the Environmental Integrity Project found that Maryland's incinerators emit more pollution—including toxics like lead and mercury—than each of the state's four large coal-fired power plants.[36] And "waste-to-energy" incineration schemes actually undermine efforts at waste prevention, since they require a constant supply of trash to stay in business. In this way, incinerators encourage increased resource extraction rather than conservation.[37]

Plastics that escape the recycle bin, landfill, or incinerator end up in the environment, where they last a very long time, polluting our oceans, harming wildlife, and making their way up the food chain. Why create disposable containers and packaging out of a material meant to last forever?

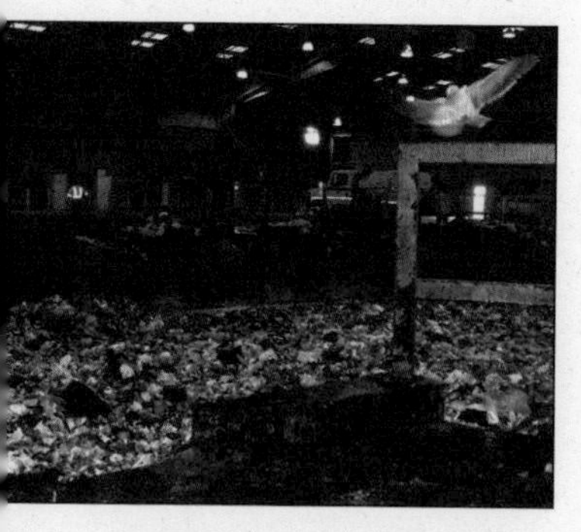

Garbage at the San Francisco dump, destined for the landfill

5. Plastic Pollutes the Ocean, Harming Both Wildlife and Humans

In 1997, while sailing his catamaran from Hawaii back to California, Captain Charles Moore, founder of the Algalita Marine Research Institute, accidentally discovered what is now known as the Great Pacific Garbage Patch. In the eastern portion of an area called the North Pacific Gyre, he was astonished to encounter enormous quantities of plastic trash floating across the sea—a garbage patch that has since been described by the media as a floating island of plastic twice the size of Texas.

But Moore and other researchers say that that description is not really accurate. The patch of plastic trash caught out in the Gyre is indeed that big, but it's more like a swirling plastic soup than an island. Plastic from land-based sources makes its way down storm drains and into waterways that eventually feed to the ocean. Currents from the North Pacific Gyre move in a swirling pattern that traps the plastic in a huge vortex, while the sun and salt water break the plastic down into smaller and smaller pieces, pieces that resemble food to hungry animals.

While no organism can digest plastic, there are quite a few animals that are harmed while trying: according to Greenpeace, at least 267 different species are known to have suffered from entanglement or ingestion of ocean plastic debris, including seabirds, turtles, seals, sea lions, whales, and fish.[38] Leatherback sea turtles choke on plastic bags they mistake for food; Laysan albatross chicks starve with their bellies full of plastic bottle caps and lighters; even the fish we eat ingest microplastic particles that have been caught in the zooplankton and pass those plastics up the food chain to us. During one trip to the North Pacific Gyre, an Algalita team found many fish full of plastic pieces, including one rainbow runner that contained eighty-four individual plastic fragments.[39]

My initial motivation for getting off my plastic diet was concern for the well-being of animals. I couldn't stand that products we used and tossed for the sake of convenience were causing harm to other creatures. But animals are not the only life forms harmed by plastic pollution. Consider this: every form of life in the ocean is connected. The food chain starts with the zooplankton, which are eaten by fish, which are eaten by even bigger fish, which are eaten by us. Plastic in fish means that the chemicals from plastic can end up on our dinner plates.

But even that's not the whole story. As Captain Moore says, plastic is both a source of pollutants (releasing its toxic additives into the ocean as it breaks down) and a sponge. Plastic is lipophilic, which means that it attracts oily chemicals. Have you

PLASTIC-FREE HEROES: Anna Cummins and Marcus Eriksen of 5 Gyres

Shortly after starting my plastic-free adventure, I received a congratulatory email from Anna Cummins from the Algalita Marine Research Institute who had also spearheaded her own campaign she dubbed "Bring Your Own" to discourage the use of disposable plastics. Anna was about to start her own adventure, one that would take her around the world and into a new relationship. She had recently met Dr. Marcus Eriksen, then a director of project development for Algalita, and the two found they had a lot in common. Marcus proposed to Anna in early 2008 during a month-long research voyage with the Algalita crew to the North Pacific Gyre. Fittingly, he fashioned a little ring for her out of trash.

That early-2008 Algalita voyage was the first leg of a campaign that Anna and Marcus called "Message in a Bottle," a way to educate the public about ocean plastic pollution. Having collected many samples of plastic-filled water, the couple sought to generate media attention and to get those samples into the hands of educators and decision makers. For the second leg—JUNKraft—Marcus and fellow adventurer Joel Paschal journeyed back out to the Pacific, from Los Angeles to Hawaii, on a raft named *JUNK*, which floated on fifteen thousand plastic bottles. Anna stayed on land and handled media and fundraising. They set out on June 4, 2008, and within four days realized the raft had started sinking—rough waves had loosened many bottle caps, and the bottles were filling up with water. Anna recounted to me the terrible phone call that began with "Baby, we're sinking." Immediately, she started calling supporters to find someone with a motorboat who would agree to rush out there. She found a boat and six volunteers and set out at 3:30 AM. At 6:00 AM, they spotted *JUNK* and spent the entire day gluing caps on bottles. Then, Anna and the volunteers left Marcus and Joel to continue on. What was supposed to be a six-week trip, ended up taking twelve. They ran out of food and had to learn to fish—and found plastic in the fish! I asked Marcus how he managed to sustain himself when the trip got hard. He said that his early military training and the importance of the message he was trying to send got him through it.

After the JUNKraft adventure, Anna and Marcus completed the Message in a Bottle campaign with the JUNKride, a 2,000-mile bike ride from Vancouver, British Columbia, to Tijuana, Mexico. All along the coast, Anna and Marcus gave presentations to public schools, universities, community groups and four city councils, giving away plastic samples from the North Pacific Gyre. They camped and stayed with local supporters. Anna says that part of the campaign was two and a half months of sheer bliss with Marcus—they even stopped along the way to get married in Big Sur, California.

I asked Anna how she and Marcus managed to fund this long adventure. She said they initially used their own savings and credit cards but had to learn to fundraise—using sites like Razoo, Facebook Causes, and Kickstarter to generate contributions from individuals and asking outdoor gear companies to sponsor them. Anna says she simply had to learn to ask for money. Once she got the first major donor on board, it became easier to ask others.

With the success of JUNKraft and JUNKride, they decided to take the project around the world, founding the 5 Gyres Institute.

The mission of the 5 Gyres Institute is to sail through all five subtropical gyres to discover the garbage patches of the world and then use their scientific work to drive change. So far, the best driver of change has been their expedition not in a gyre, but in the Great Lakes. They found abundant microplastics in the form of spherical microbeads that matched the same microbeads in over-the-counter facial scrubs. (You'll read more about those in Chapter 7.) The study was published,[40] the public signed petitions, and soon companies were announcing their plastic microbead phaseouts. It was a great success!

Today, 5 Gyres invites members of the public to join its research missions to see the problem up close for themselves. Some day I hope to join such an expedition! For now, I settle for the small, glass jar of ocean plastic that Anna brought me from the Great Pacific Garbage Patch. It reminds me not only of the problem, but also that there are inspiring, committed activists out there willing to take risks to make the world a better place, that these activists are regular people like you and me, and that activism can be fun!

ever noticed how hard it is to clean a plastic container that has contained greasy or oily food? Sadly, the ocean is home to oil-based pollutants that run off from land, chemicals like PCBs and DDT, which are even more harmful for us to ingest than the plastic itself. So because of the way plastic attracts oil, plastic particles in the ocean are often coated with these toxic oil-based pollutants. In fact, ocean researchers from Tokyo University have found concentrations of these chemicals a million times higher on plastic particles than in the surrounding sea water. So we get an extra serving of chemicals with our plastic fish. Just something to think about.

Something else to consider is that the problem of plastics in the marine environment is not confined to the Northern Pacific Ocean. There are actually five major gyres in the world's oceans: the North Pacific, South Pacific, North Atlantic, South Atlantic, and Indian Ocean gyres. During a recent voyage to the South Atlantic Gyre, researchers from the 5 Gyres Project found plastic in every water sample they collected.

With new plastic flowing into the world's oceans every minute, ideas about cleaning up the Gyres are impractical. You might hear stories about plans and projects aimed at recovering and recycling the plastic trash from the ocean. And these projects might very well be useful. But there's no way we will ever clean up all of it as long as we continue to consume new plastic at our current rate. As Captain Moore described it on a recent David Letterman show, "It would be like sifting the Sahara Desert." Or bailing water from a bathtub while the spigot is on.

I believe we all have the power to turn that spigot off.

Sample of plastic from the North Pacific Gyre taken by the researchers from the Algalita Marine Research Foundation in 2008

Ocean Plastic Pollution Resources Here are a few educational resources with more information about ocean plastic pollution and harm to wildlife:

- ***Midway Journey: Message from the Gyre*** This film chronicles the journey of photographer Chris Jordan along with a poet, an activist, and several filmmakers to document the effects of plastic pollution on the albatross population on Midway Island.

• ***Chris Jordan's Midway: Message from the Gyre*** A series of beautiful yet heartbreaking photos of albatross chick carcasses filled with plastic pieces.

• ***TEDx Great Pacific Garbage Patch*** The Plastic Pollution Coalition, in conjunction with the TED organization, presents inspiring videos from many experts and activists on solving the plastic pollution problem. I gave my first talk on plastic-free living during this event. All videos are available on the TEDx YouTube channel.

Books

• ***Plastic Ocean: How a Sea Captain's Chance Discovery Launched a Determined Quest to Save the Oceans,*** by Charles Moore. Avery, 2011. Captain Moore's amazing adventure in discovering the Great Pacific Garbage Patch, told in his own words.

• ***Plastiki: Across the Pacific on Plastic: An Adventure to Save Our Oceans,*** by David de Rothschild. Chronicle Books, 2011. Explorer and eco-TV host David de Rothschild recounts the journey of the *Plastiki*, a catamaran built from recycled plastic, to bring awareness to the plight of the oceans.

• ***Sullie Saves the Seas,*** by Goffinet McLaren. Prose Press, 2011. Children's book about how a seagull calls his ocean friends to action to combat plastic pollution.

6. Plastic Pollution Doesn't Just Come from "Litter Bugs"

The plastics industry likes to insist that plastic pollution is the fault of careless consumers who don't dispose of their waste properly. And I have certainly seen my share of kids walking down the street leaving behind a trail of trash or drivers casually tossing cigarette butts (which are actually plastic) or other garbage out the window. But these "litter bugs" are not the only source of plastic litter.

I've also seen the way the trash trucks come through my neighborhood with their mechanical arms lifting our garbage cans and dumping the contents into the back. Often they'll leave plastic wrappers and bags blowing down the sidewalk afterwards, either because the truck mechanism is not entirely accurate or because lightweight plastics blow out on windy days. I've seen waste cans in parks overflowing with garbage because the city's financial resources are too strained to allow for frequent pickups.

And I've seen plenty of people, myself included, accidentally drop something and keep walking rather than turn around and pick it up. But these are all sources of "post-consumer" litter—litter consisting of the trash left after we've consumed the contents of the package, worn out the flip flop, or broken the sunglasses. There is another source of plastic litter the industry doesn't like to admit to:

Overflowing public trash cans

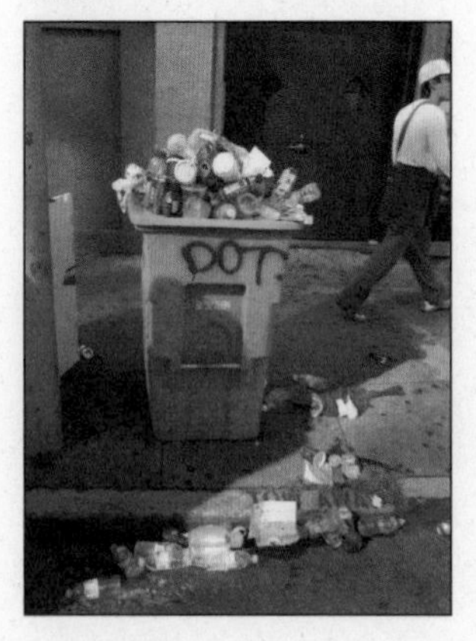

Nurdles.

Who? What? "Nurdle" is the nickname for the tiny pre-production plastic pellets that are the basis for all plastic products. They look like tiny, white fish eggs, which is a problem for animals that actually eat fish eggs when nurdles get loose in the ocean. Companies that manufacture raw plastics sell their product in the form of tiny resin pellets to the companies that manufacture the actual bottles, utensils, shower curtains, lunch boxes, cups, toothpaste tubes, or other plastic products that we use every day. The problem is that plastic nurdles can blow away just as easily as plastic bags or wrappers. And they do.

Nurdles have been found littering the ground around the buildings of plastics manufacturing companies or along railroad tracks after being spilled in transit. And they are a huge component of the plastic litter on the world's beaches. I've seen for myself nurdles washed up along a beach in Marin County, California, where the plastic pellets just blend in with the sand if you're not looking for them. In fact, the name "nurdle" was supposedly coined by surfers who would pick them up from the sand and chew on them, not knowing what they actually were.

The problem of plastic resin pellet pollution in the ocean has become so prevalent that California passed legislation in 2007 ordering the State Water Resources Control Board to implement a program for the control of discharges of preproduction plastics, but how much can one state actually do when ocean plastic pollution from resin pellets is a global problem?

Nurdles collected from Kehoe Beach in Marin County, California, in spring 2010

To its credit, the plastics industry responded to the problem of resin pellet pollution with their Operation Clean Sweep Initiative, a campaign which asks plastics manufacturers to "pledge to prevent pellet loss" by following a set of guidelines and procedures designed to keep resin pellets out of the ocean. It's a good step, but unfortunately, the guidelines have no teeth. The program is completely voluntary. In fact, the first paragraph of the OCS manual states, "You are encouraged to implement the sections and steps that help achieve your company's specific goals. ***None of the guidelines are intended as a mandate.***"[41] So whether Operation Clean Sweep is more than just good PR remains to be seen.

Brand-new nurdles at plastic bag factory in Northern California

The point is that no matter how careful we are to put our own plastic wrapper in the proper receptacle, there is no guarantee that it will end up where we want it to. And even if it does, the company that produced it might have already been the source of preproduction plastic litter before that wrapper ever got to us. All of us who use plastic support an industry that produces plastic litter. And all of us have the power to cut off our support. It's as simple as that.

Take action: Clean up a beach or even a creek. Going outside to help clean up a beach, river bank, creek, or other body of water is a good way to understand the impact of plastic trash. While we can't solve the plastic pollution problem by cleaning up a section of beach once a year, doing so can be very educational. It's a great activity to do with kids. Simply picking up trash from around the neighborhood is also a good way to get started. You might not live near a body of water, but you can bet that any litter on the sidewalk will make its way to storm drains during a good rain, into waterways and eventually out to sea. Unless you live below sea level, the ocean is downhill from you. In addition to the Algalita Marine Research Institute and 5 Gyres, check out these

Plastic trash litters Oakland's Damon Slough before making its way out to the San Francisco Bay and Pacific Ocean.

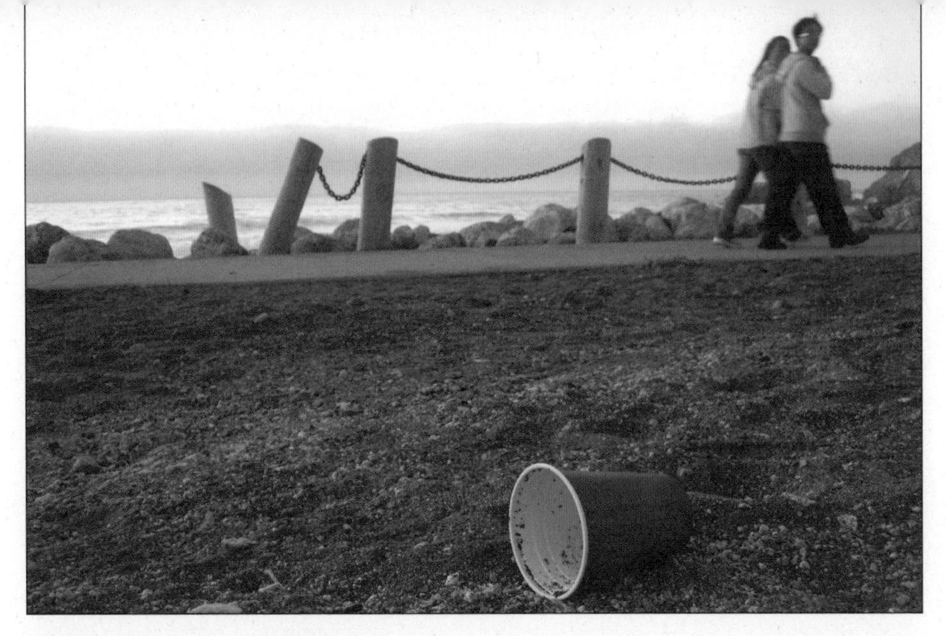

Beach trash: polystyrene plastic cup

other inspiring organizations and individuals working for waterways free from plastic pollution:

- **The Ocean Conservancy** organizes an International Coastal Cleanup Day every fall. Visit the website to sign up for email updates and find a cleanup event near you.
- **American Rivers** organizes the National River Cleanup, an ongoing initiative that helps organizers and volunteers come together to clean up America's rivers. Find a river cleanup or register a new event on the website.
- **The Daily Ocean** is the blog of ocean lover Sara Bayles who committed to cleaning up a year's worth of trash from a section of beach in Santa Monica, California. After collecting over 1,300 pounds of trash, she's still going. Sara posts pictures of her finds, invites other locals to join

her, and speaks about her project to schools and businesses. Find her at dailyocean.org.

- **It Starts with Me** (www.itstartswithme-danielle.blogspot.com) is the blog of North Carolinian Danielle Richardet who takes her kids out to the beach to collect trash. Danielle has found that the most common type of litter is cigarette butts, a type of plastic. In fact, she teamed up with the Brita Company to make a short movie about her project, which you can also view on her site.
- **The Plastic Ocean Project** was founded by another North Carolinian, Bonnie Monteleone, who combines art with collaborative ocean research and education to incubate solutions to address the global plastic pollution problem.
- **Plastic Free Seas** is a Hong Kong based environmental organization founded by Australian Tracy Read, who in 2012, discovered one of the world's largest documented plastic pellet spills and began a tireless campaign to hold the responsible parties accountable and get the mess cleaned up. Plastic Free Seas advocates change in the way we all view and use plastics in society today, through education and action campaigns.
- **Two Hands Project,** founded by Australian Paul Sharp, is a collaborative approach to dealing with plastic pollution: Take 30 Minutes and Two Hands to clean up your world anytime, anywhere.
- **Take 3 Clean Beach Initiative,** founded by yet another Aussie, Tim Silverwood, has a simple message: take 3 pieces of rubbish with you when you leave the beach, waterway, or anywhere and you have made a difference.

7. Plastic Recycling Doesn't Close the Loop

Most plastic recycling is actually downcycling. Plastic bottles are rarely recycled into new plastic bottles but into other products such as carpet or polar fleece, which means that the recycling loop isn't closed. Downcycling a soda bottle into a carpet means that virgin plastic must be used to create the next new bottle. The process doesn't reduce production of new plastic, which should be the intent of real recycling.

I've got a lot more to say about plastic "recycling" and what actually happens to it when it leaves your recycle bin. In fact, I've got so much to say that I've given the topic its own chapter. For now though, it's enough to know that recycling plastic is problematic, and while some plastic recycling is necessary, it doesn't get us off the hook.

Types of Plastic: The Numbers on the Bottom

Throughout my discussion of the problems with plastic, I threw out some numbers and names. And you've probably seen them yourself: those tiny numbers inside the "chasing arrows" that look like a recycling symbol on the bottom of a bottle or container. They appear on a handful of items in my plastic stash—drink bottles, a vitamin bottle, a squeeze bottle, a soup container, a hummus tub, and a clamshell container. Those numbers are called the "SPI resin identification code" (also known as the material container code.) And before I explain what those numbers mean, I need to tell you what they *don't* mean:

1. A Number Inside a Chasing Arrows Symbol on a Plastic Container Does *Not* Mean that the Container Is Reyclable. You heard me. That symbol does not mean the container is recyclable.

When I first started this project, I was eager to learn what plastic numbers my city accepted and get those items into the recycle bin where they belonged. But contrary to what I, like most people, thought, that symbol merely indicates the type of polymer the container or bottle is made from. Recyclers can use the code to help them sort plastics. But the symbol doesn't tell you whether the object can actually be recycled where you live or whether it's recyclable anywhere at all! In fact, the SPI (Society of the Plastics Industry) requires manufacturers who use the symbol to mold it unobtrusively onto the very bottom of the container or bottle and prohibits them from using the word "Recyclable" on the label within close proximity to that symbol.

Why is the plastics industry concerned about its codes being used in any other manner? According to SPI, "alleged abuses of resin identification codes have led consumer and environmental groups to ask the Federal Trade Commission (FTC) and State Attorneys General, among others, to take legal or regulatory action." Indeed, the FTC has published *Guides for the use of Environmental Marketing Claims* that, among other things, prescribes how claims of recyclability should be stated to avoid deception. Knowing that many consumers believe the chasing arrows symbol indicates a product's recyclability, the FTC itself directs plastics manufacturers to place the resin identification code in an inconspicuous location on the container so as not to imply that the container can be recycled.

Unfortunately, the FTC's *Guides for the use of Environmental Marketing Claims* are just that: guides. They do not carry the force and effect of law. So consumers have to be very careful and ask a lot of questions when it comes to environmental claims. And believe me, since I started looking into the problems with plastic, I've asked a lot of questions! I'll share with you what I've learned and what types of questions to ask in future chapters. But for now, we're just talking about those resin codes, and what they don't tell us.

2. The Resin Identification Code Does Not Tell You If a Product Is Safe! The SPI code only tells you what the polymer is, but not what other chemicals may have been added to that polymer. As I mentioned before, knowing what type of plastic something is made from does not mean we know what else has been added to it.

3. The Code Numbers 1 Through 7 Do Not Cover Every Type of Plastic. The resin identification codes are a way for recyclers to quickly determine what type of disposable plastic they are dealing with. Therefore, SPI only created codes for the plastics most commonly used in disposable plastic packaging and containers—the types of plastics most likely to end up in the recycle bin. Numbers 1 through 6 indicate specific types of plastic, and number 7 is a catchall for "everything else." The majority of the plastic I have collected, whether numbered or not, has been "everything else."

Now that I've gotten those caveats out of the way, here is a handy chart explaining the SPI resin codes one by one, what types of plastic they indicate, what types of products are made from those plastics, and the problems associated with each kind.

SPI Resin Codes Identification Chart[42]

1 PETE	**Polyethylene Terephthalate (PET, PETE).** PET is clear plastic most commonly used in bottles for sodas, water, juice, and other drinks. Jars for peanut butter and other foods. Microwavable dinner containers. Polyester fabric. PET bottles are recycled into secondary products like polar fleece and carpets.
2 HDPE	**High Density Polyethylene (HDPE).** HDPE is heavy opaque or translucent plastic used in milk and water jugs, bottles for shampoo, laundry detergent, and household cleaners, as well as grocery bags, cereal box liners, some medicine bottles, plumbing pipes, and plastic/wood composites. HDPE bottles and jugs are recycled into bottles for non-food items, plastic lumber, pipe, floor tiles, buckets, crates, flower pots, garden edging, film and sheet, and recycling bins. HDPE grocery bags are most often recycled into plastic/wood decking material.
3 V	**Polyvinyl Chloride (PVC, Vinyl).** Rigid PVC is used for pipes, vinyl siding, window frames, fencing and decking, as well as blister packs and clamshells. When PVC is softened with plasticizers, it is used for cling wraps, squeeze bottles, soft toys, loose-leaf binders, lunch boxes, blood bags and medical tubing, wire and cable insulation, carpet backing, and flooring, and many other products. Hazards of PVC: dioxin, a known carcinogen, is a by-product of manufacture. Toxic additives like lead and phthalates are often added to PVC. PVC is rarely recycled.
4 LDPE	**Low Density Polyethylene (LDPE) and Linear Low Density Polyethylene** **(LLDPE).** Mostly used in film applications such as bags for dry cleaning, newspapers, bread, and garbage bags. Some shrink wrap and stretch film. Coatings for paper milk cartons and beverage cups. Also some squeeze bottles, container lids, and toys. LDPE films can be hard to recycle.

5 PP	**Polypropylene (PP).** Rigid plastic used for yogurt, margarine, and cottage cheese tubs, some medicine bottles, bottle caps, Brita water filter cartridges, and durable consumer products like reusable food storage containers, sippy cups, appliances and automotive applications. PP can be difficult to recycle, but several dedicated programs exist such as Preserve's Gimme5 Program. PP foodware often contains antibacterial chemicals.
6 PS	**Polystyrene (PS).** Rigid PS can be clear or opaque and is used for clamshells and a variety of disposable foodware such as plastic drink cups, plates, and utensils. CD and DVD cases. Many other consumer products. Foamed PS, often known by the brand name Styrofoam, is used for packing peanuts, take-out foodware, and packaging for electronics and furniture. PS is very difficult to recycle. PS contains styrene, a suspected carcinogen.
7 OTHER	**Other.** This category includes all other plastics not listed above, including: **Polycarbonate (PC).** Rigid plastic used in sports bottles, 5-gallon jugs, baby bottles, CDs and DVDs, food processors and juicers, eyeglass lenses, and other products. PC contains Bisphenol-A, an endocrine disruptor linked to many health problems. **Bioplastics.** Plastics like Polylactic Acid (PLA) and Polyhydroxyalkanoate (PHA) that are made from plants rather than fossil fuels are also labeled as #7. Some bioplastics are compostable in the right conditions. **Composites.** Mixtures or layers of different types of plastic.

Besides the handful of plastics listed in the SPI Resin Codes Identification Chart, there are many more which are never marked with a code, including nylons, acrylics (Plexiglas), epoxies, polyurethanes (foam rubber, varnishes, and adhesives), other synthetic rubbers (Neoprene, EDPM), phenolics (Bakelite, Formica), fluoropolymers (Teflon), and many, many more.[43]

Check these additional resources for more information about plastic pollution and what you can do to be part of the solution:

Organizations

- **Plastic Pollution Coalition** (PPC) is a global alliance of individuals, organizations, and businesses working together to stop plastic pollution and its toxic impacts on humans, animals, and the environment. Campaigns include REFUSE!, a global campaign to encourage people to stop using disposable plastic; Plasticfree Towns; Plastic-free Campuses; art and artist events; and film, video, and social media initiatives. Become a member and sign the pledge to refuse single use plastics.
- **Green Sangha** is a Bay Area organization based on principles of nonduality and shared oneness that brings spiritual practice and environmental activism together. The Rethinking Plastics campaign advocates for community zero waste and plastics reduction measures and provides in-depth workshops on the history of and issues related to plastic, shorter presentations for schools, organizations, and community groups, and a travelling display.
- **Kokua Hawai'i Foundation** supports environmental education in the schools and communities of Hawai'i, including Plastic Free Hawai'i, a coalition of community members and business owners that strives to educate the stores, schools, restaurants, residents, and visitors of Hawai'i on the environmental and health benefits of going plastic-free; Plastic Free Schools Hawai'i, which includes an educators' resource guide; and the Waste Free Lunches program.

Films

- ***Addicted to Plastic*** (2008)
- ***Plastic Shores*** (2012)
- ***Plastic Paradise: The Great Pacific Garbage Patch*** (2013)

Take the "Show Your Plastic" Challenge

I learned a lot about myself and my personal habits by collecting my plastic waste and analyzing it each week. As I said, the exercise was like a game. So in the spring of 2009, I threw down the gauntlet and invited my blog readers to join in the fun. And now, I'm inviting you too.

Understanding what your own personal plastic waste profile looks like will give you a reference point as you go through the next sections of the book.

Here's how it works:

1. Collect All of Your Own Plastic Waste, both recyclable and non-, for a minimum of one week. If you want to go for more than one week, great! Just keep each week's collection separate.

2. What Qualifies as Yours? Anything that benefits you. So, if your housemate or significant other brings home a tub of yogurt that you both share, the tub goes in your tally. But if you hate yogurt, never touch the stuff, and wouldn't have bought it for yourself in a million years, it's not your responsibility. What about stuff for your kids? I'll leave that up to you. Whatever you decide, just be consistent about collecting it.

3. Live Normally in the First Week. It doesn't help to artificially reduce your plastic consumption for the sake of a one-week tally if you will go back to living with more plastic afterward. Think of this exercise as a scientific experiment. Nothing more.

4. List Out the Items at the end of each week that you do the exercise. You might also want to take a photo. Including details about what things are recyclable in your community can be helpful. If you have a sensitive food scale, you might consider weighing your plastic as well. A bathroom scale won't really be accurate because plastic is so lightweight.

5. Guilt is Not Encouraged. Nor are comparisons with other people whom you perceive to be doing "worse" or "better" than you in terms of plastic waste. This exercise is for purely educational purposes. Guilt doesn't help.

6. Answer the Questions Listed on the Worksheet on the Next Page. As you're starting out, you may not have answers for many of the questions. That's okay. As you go through the book, you'll find solutions and plastic-free alternatives to the items in your tally. But having this tally as a baseline can help show you what areas to focus on. You can always come back to the worksheet and fill in answers as you discover them later.

7. Optional: Go Public. If you'd like to post your photo and challenge results to the Show Your Plastic Trash Challenge website (a great way to get feedback from other participants and blog readers), visit www.myplasticfreelife.com/showyourplastic/.

Show Your Plastic Trash Challenge Worksheet
Name: ____________________ Week Ending Date: ____________________ Total number of items: ____________________ Total weight of plastic stash (If you're able to do this): __________ List of Recyclable Items: (Write out in list form and include the SPI resin code and how the item gets recycled in your community, as far as you know. If you are not sure if any of your items are recyclable, leave this field blank and enter all your items in the next box.)
List of non-recyclable items or those you are unsure of:

Looking at my photo and list, what feelings arise for me?
What items could I easily replace with plastic-free or less-plastic alternatives?
What items would I be willing to give up if a plastic-free alternative doesn't exist?
What items are essential and seem to have no plastic-free alternative?
What lifestyle change(s) might be necessary to reduce my plastic consumption?
What one plastic item am I willing to give up or replace this week?

What other conclusions, if any, can I draw?

GIVING OURSELVES A BREAK, A FEW WORDS OF INSPIRATION

People often tell me they resist the Show Your Plastic Challenge because of personal guilt. They are afraid they will feel bad about what they discover. I tell them guilt doesn't help and that it's important to do this project objectively without blaming ourselves or beating ourselves up. But honestly? I beat myself up sometimes too. So I decided to ask my meditation teacher, Jon Bernie, if he had any advice to share with us. Jon is highly sought after in the Bay Area for his depth of compassion and clarity. His semiannual meditation retreats keep me sane throughout the year. I figured if anyone had some good ideas about getting past guilt to understanding, it would be Jon. Here's what he said:

> Beth, you and I are both asking people to become conscious. Your work is specific in terms of actions and changes in behavior to protect the environment. My work is about inner transformation, inner ecology. If people are worried about doing the exercise because they think they might be overwhelmed by the pile of plastic trash that they end up looking at the end of the week, it's really important to

acknowledge that. First, be willing to feel whatever it is that you're feeling. The emotion that arises is actually just another river that needs to flow. It's as much a part of the eco-system as anything else, and if we block that, we're basically trashing ourselves. The more we relax and get out of the way, the better that energy can move. Pay attention to the physical sensations in the body, the breath, the belly. Maybe you're afraid. Maybe you cry. It's okay. The mind may try to defend itself with stories of guilt or blame, but that's just a defense. Ask yourself if those judgments and beliefs are true.

If one can separate the actual feeling from the beliefs and conclusions about the feeling and just allow the fear, guilt, anger, whatever it is to move, then we've created some space to look at that pile of plastic trash with more clarity. We can look at our habits, we can look at the plastic, we can look at those things and go, 'Wow, gee, do I really need that and that and that? No maybe not. But I do need this and this.' We can assess our plastic use from a kind of balanced, neutral place where we're not overwhelmed. Recycling begins in one's own heart, and if one isn't recycling one's own garbage internally, it ain't gonna happen on the outside. —Jon Bernie, healer/spiritual teacher, Clear Water Sangha, www.jonbernie.org

Wow. All this time I'd been telling people just to not feel guilty. But the truth is that if we're going to look at our plastic use with objectivity, we have to be honest about whatever emotions arise for us in the process. And just as we question our habits and purchases and patterns of consumption, we can question our beliefs and judgments about ourselves and the world. These are lessons that I learn—and relearn—over and over again, sometimes to comic effect. Keep reading. You'll see what I mean.

Chapter 2: Plastic Bags (Why Are There Melons in My Shirt?)

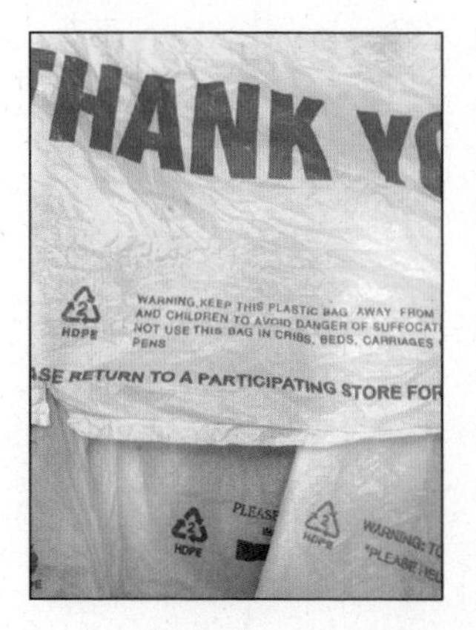

If there's one thing all of us can do to reduce the amount of plastic waste we generate, it's say no thanks to disposable plastic shopping bags. Back in 2007, looking at the pile of waste I generated in my first week of collecting plastic, I could see it was going to take a while to replace all the wrappers and containers and packaging with plastic-free versions. But the one change I was pretty sure I could make immediately was to start bringing my own reusable bags with me to the store and to stop accepting disposable plastic ones once and for all. That step would be easy. Right?

I scrounged through my closets for the free tote bags I'd picked up at various events and never bothered to use and put them in my backpack to bring to the store. And just to be sure I wouldn't get caught without a bag in the future, I dragged out some old plastic grocery bags from the bin under the sink and stuffed them into various purses and jacket pockets. I figured, better to reuse what I already had than to get caught unprepared. And then, self-consciously, I wrote the words REUSABLE BAG in thick black marker on one of the plastic grocery bags, so no one would mistake my repurposed plastic bags for new ones. I was feeling pretty self-righteous.

That smugness melted an hour later when I found myself in line at the grocery store waiting to check out. It seemed that I was the only one with reusable bags. I watched as one customer after another in front of me accepted their groceries in plastic bags and hurried away. There went a gallon jug of milk in its own plastic bag, even though it already had a handle. There went a plastic produce bag of apples into another plastic grocery bag. And there went the guy right before me, dropping the plastic bag he'd just been given into the recycle bin outside the store after removing his one pack of gum and bottle of soda. Why had he taken the plastic bag in the first place?

Then it was my turn. I was starting to feel like a freak. Everyone was in such a hurry. No one else was making special requests. And here I was with my cloth bags getting ready to hold up the line. I imagined the customers behind me rolling their eyes and looking at their watches. My face felt hot as I stammered too quietly, "I have my own bags." Without looking up, the cashier started to put my groceries in a plastic bag. "Wait. No. I have my own bags," I repeated a little louder, holding them up to show her. Sighing, she took my purchases out of the plastic bag, then balled up the bag and started to throw it away.

"Wait! What are you doing?" I protested. "You can use that bag for someone else!" Shrugging, she tossed the bag aside and continued scanning my purchases, leaving each item on the counter for me to bag myself. Apparently, bringing my own bag meant bagging my own groceries. Not wanting to make more of a fuss, I hurried to pay for my purchases and then stood at the end of the counter bagging my own stuff while the cashier rang up the next customer. I felt alone and embarrassed. And then I felt angry. Here I was trying to do the right thing and being made to feel like a weirdo for giving a crap about the planet. I wanted to yell at each person passively accepting a plastic bag, "Wake up! Don't you understand what you're doing?"

Walking home that afternoon, I replayed the experience in my mind. Was it true that people thought I was weird? Was it true that I had actually even held up the line? I realized that the only person making me feel embarrassed was me. The people in that grocery line had no more idea of what was going on in my head than I did what was happening in theirs. So why not spin a more positive story? Maybe there were people in that line that noticed me with my bags and made a mental note to remember to bring their reusable bags next time. Maybe the cashier thought to herself, "Next time I'll ask if the customer wants a bag or not." Maybe someone else back in that line did bring his reusable bags and seeing me use mine helped him feel part of a movement bigger than himself.

The truth is I have no idea what any of the people in the line that day were thinking. What I do know is that each time throughout my plastic-free experiment I have had to speak up or try something new, I've reminded myself of that first time bringing my reusable bags to the grocery store, how strange it felt at first, but how I did it anyway. And I remember how with practice, it got easier and easier and felt more and more normal. Nowadays, bringing our own bags to the store is pretty commonplace. Many stores give a discount for refusing disposable bags. And even stores like Wal-Mart with automated self-checkout options have programmed their machines to ask customers if they have their own bag. Bringing our own bags to other types of stores—pharmacies, department, clothing, etc.—is a little less common, but there's no reason it shouldn't become the norm the more of us do it. It just requires the willingness to speak up with polite conviction. A smile goes a long way, too.

What's Wrong with Plastic Bags?

Aside from all the issues with plastic I listed in chapter 1, plastic bags have some unique problems. While their environmental costs are burdensome for communities and the planet, the cost of plastic bags for retailers is pretty low. Made from ethylene, a byproduct of petroleum or natural gas, plastic bags are so cheap and flimsy that cashiers use them freely, double bagging as a matter of course and often sticking just a few items in each bag. As a result, shoppers end up with piles of plastic bags spilling out of closets and threatening to take over cupboards . . . until we finally throw up our hands and either dump them in the trash or, if we're lucky enough to live in an area where stores provide plastic bag collection bins, cart them back for recycling. Sure, some of us reuse plastic shopping bags to line our waste bins or to pick up dog poop, but the bags still end up in the landfill. (I'll discuss ways to handle the garbage/poop bag problem in chapter 7.)

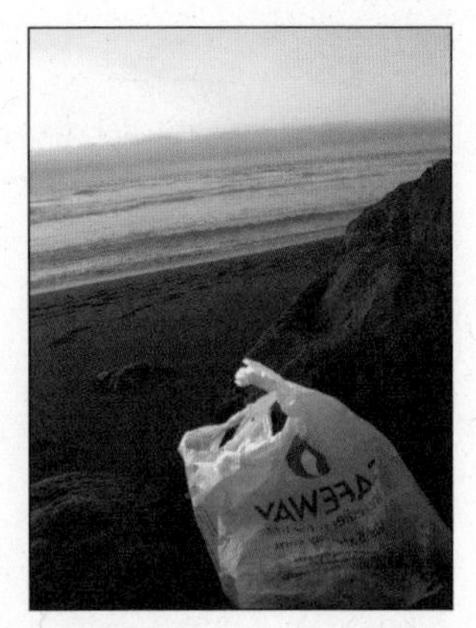

A plastic grocery bag litters the beach.

Even when disposed of properly, plastic bags are so lightweight and aerodynamic, they are easily picked up and carried by the wind. They can escape from trash bins, recycle bins, garbage trucks, and landfills, and end up littering the landscape. Blowing down the street, flapping from trees, clogging storm drains (costing municipalities millions of dollars in cleanup costs), and making their way out to sea, plastic bags have been referred to as "urban tumbleweeds" for good reason. And they persist in the environment, causing harm for a very long time.

Eating Plastic Bags for Lunch

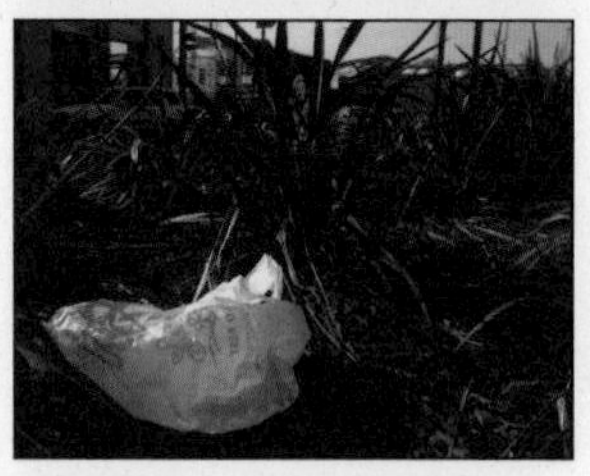

Urban tumbleweeds in San Francisco

A Google search on "animals eat plastic bags" brings up hundreds of heartbreaking stories and images from around the world. So many foraging cows in India have died from ingesting plastic bag litter that many of the states in that country have banned the distribution of plastic bags. In the United Arab Emirates, a veterinarian has documented images of camels, sheep, goats, and endangered desert animals dead from eating plastic bags.[44] Whales wash up on our coasts, their bellies full of plastic.[45] And endangered leatherback sea turtles mistake floating plastic bags for the jellyfish that are their main diet, ingesting the plastic that can then block their digestive tracts. In fact, a recent study of leatherback turtle autopsy records found plastic in one-third of the animals' GI tracts, plastic bags being the most common item mentioned.[46]

I wondered just how many marine animals died from eating plastic bags each year, so I called my friend Wallace "J." Nichols, a marine biologist and research associate for the California Academy of Sciences. J. told me that while a lot of figures have been thrown around in the media, hard numbers are difficult to calculate, and the sad fact is that when most sea animals eat plastic and die, they sink to the bottom, unaccounted for. But possibly more significant than the individual animals that are killed by eating plastic are those that are affected indirectly. For example, when sea turtles eat plastic instead of food, their glucose levels drop, leaving them with less energy for migration and reproduction. Females can't lay as many eggs, and fewer new sea turtles are born. "When you connect the dots," J. said, "you realize that plastic pollution may cost millions of potential sea turtle lives."

I asked J. if he'd actually seen a sea turtle eat a plastic bag, and he shared the following story:

> Once we were out on the water, and there was a turtle just floating. Usually when you get close to them, they dive and swim away, but this turtle didn't. We thought maybe it was tangled in something like a net. But the closer we got, the more we realized it wasn't tangled in anything. It was just incredibly bloated, like it was wearing a big flotation device, except the flotation device was inside it. We put the turtle in the boat because a turtle left in that condition is just going to die. It can't dive to eat. It's vulnerable. It will starve. We got it back to the lab and into a pool and observed it for a while. And then we realized it had something blocking its gut, and that a piece of that something was hanging out.

At that point J. told me he needed to step outside to tell me the rest of the story because he didn't want to gross out the people around him. I won't gross you out with the details either. Suffice it to say that what was hanging out of that turtle's butt was a plastic bag, and once they got that bag out, plus all the stuff that was backed up behind it, that turtle felt a whole lot better. "There was a collective sigh of relief around that turtle. Everyone was empathizing. It was like giving birth to a plastic bag."[47]

Sad stories about turtles, whales, and camels break my heart. But for me, the problem hits even closer to home: My cat eats plastic too.

Call it ironic, but my little cat, Arya, loves the stuff. She ate big holes in two polyester fleece blankets and pooped red fuzz for several days. She once swallowed and coughed back up a string cheese wrapper (not mine!) mixed with a hairball. Those cases were bad enough, but a few months after the string cheese incident, Arya got very sick. She developed diarrhea, wouldn't eat, and vomited constantly in every room of our apartment. We took her to the vet and discovered she had a fever. We spent over $200 on tests and medication for her, but the blood tests were inconclusive and the medicine didn't seem to help. Michael and I were worried as hell. Until one morning, lying pitifully in the middle of the bedroom floor, Arya vomited up the final hairball. It was full of—you guessed it—a big chunk of plastic bag.

Apparently, Arya is not unique. I'm not sure whether it's the sound, smell, or texture, but for whatever reason, many cats and dogs think that plastic bags are food and will gnaw, lick, or suck on them every chance they get. Try Googling, "Why does my cat eat plastic bags?" and you will find page after page of results. Could our pets be attracted to the animal fat that is added to some plastic bags to make them slippery? Or some other mystery additive? Whatever the reason, our companions rely on us to protect them. But who protects wild animals trying to cope with an increasingly polluted environment that they didn't choose? It's up to us as individuals and communities to refuse single use plastic bags.

Are Paper Bags Better?

Only marginally. But in the beginning of my plastic-free experiment, I assumed, like many other people, that paper bags were the eco-friendly alternative to plastic. After all, paper bags are biodegradable. Loose in the environment, they fall apart very quickly, especially after a good rainstorm. You don't hear about animals becoming entangled in paper or dying with bellies full of the stuff. And paper bags come from a renewable resource—trees—which can be planted over and over again in sustainably managed forests. If we could just get stores to switch back to paper bags, our problem would be solved, right?

Not so fast. It turns out that transforming trees into paper is an incredibly energy and water intensive process. In fact, some life-cycle analyses have calculated that paper bags require three times more energy to produce and ship than plastic bags, despite the fact that the raw material for plastic bags comes from fossil fuels. What's more, paper bag manufacturing uses an entire gallon of water per paper bag compared with .008 gallons per plastic bag.[48] And what about compostable bags? They have their environmental impact too. (I'll talk more about compostable bio-based plastics in chapter 5.)

Organic cotton grocery bag

The conclusion? All disposable products have an environmental cost. Given the rare emergency situation in which my only options are paper or plastic, I would accept a paper bag and then reuse it as many times as possible, knowing that at the end of its useful life, whether recycled or composted, it won't linger in the environment to harm animals or clog storm drains. But the real solution to the plastic bag problem is to reduce our consumption of single-use disposables in general, a theme that you'll find repeated throughout this book.

Reusable Bags—They're Not All Created Equal

As a plastic-free blogger, I receive emails every week from PR reps who would like me to review the latest and greatest reusable bags. Made from canvas, net, hemp, organic cotton, recycled soda bottles, polyester, polypropylene, burlap, nylon, and accessorized with pockets, stuff sacks, carabiners, drawstrings, and other accoutrements, reusable bags vary not only in brand and style, but also quality, price, and environmental impact.

Reused/Repurposed Bags

Stocking up on reusable shopping bags doesn't have to be expensive. Repurposing materials we already have keeps costs down and creates the least environmental impact.

The Bags You Already Have The cheapest and greenest bags are those that you already own. Do you have a closet full of unused tote bags like I did? Dig them out.

Or broaden your concept of what a shopping bag is in the first place. It could be a backpack. Or a purse. Duffel bag. A wicker basket or even the removable wire basket from your bike. In fact, my amazing friend Tracey once used a wheelie suitcase as a shopping bag/cart. You'll learn more about her in chapter 6.

Reused Plastic Bags Is the Bag Monster alive in the cupboard under your kitchen sink? Set it free. Advocates for the plastic bag industry argue that disposable bags are actually reusable, and they are right . . . to a point. Disposable bags are thin and cheap and don't last long, but they can be reused for their original purpose until they wear out. Keep a few plastic bags in your purse, backpack, pockets, or wallet to avoid being caught bagless. Just be sure and rinse out dirty bags between uses. We hang wet bags on a wooden Bag-E-Wash bag dryer over our drain board. Other people hang plastic bags over wooden chopsticks or long utensils. The key is to make sure the bag stays open upside down while draining to avoid growing mildew.

Secondhand Bags from Thrift Stores, Freecycle, Etc. With the popularity of the Bring-Your-Own-Bag movement, companies have latched onto the idea for promotional purposes. Anyone who goes to conferences probably has a pile of reusable bags lying around sporting company logos, and based on the number of reusable bags I saw left behind in the "swag recycling room" at the last conference I attended, many of us have more than we need. Freecycle.org is a great resource for matching things one person doesn't want with someone who actually needs it. Why not post an ad asking for reusable tote bags or used plastic bags? Or check out your local thrift stores or Craigslist.org for used bags. And if you're someone who has too many bags, consider spreading the wealth. Michael likes to carry extra reusable bags with him and offer them to shoppers in line who might have forgotten to bring theirs. In fact, the night I wrote this chapter, he came home and reported that he'd helped out a woman on a bicycle whose disposable grocery bags were threatening to fall apart before she got home. He carried her groceries the few blocks to our house and then gave her a couple of extra bags we had lying around. What a guy!

> "My golden rule is to never buy reusable bags. In the beginning, I didn't want to create more waste, so I just got a few old backpacks and totes to bring to the market. Nowadays, every store and brand gives away reusable bags, but I only accept them if they're plastic-free and I forgot my own at home. Whenever I feel like I have too many reusable bags, I just give a few to someone who I know is trying to quit plastic bags."
> —Yessica Curiel Montoya, New York, New York

Do-It-Yourself Bags from Old T-Shirts and Other Repurposed Materials A Google search on "shopping bag pattern" brings up a long list of instructions, patterns, and ideas for DIYers. I've provided a simple pattern on page 64 for creating a reusable bag from an old t-shirt. Or check out the YouTube video demonstrating how artist Ame Guzeman is able to create five bags out of one shirt (On Youtube, search for "Zero Waste Project Baygs"). If you're not particularly crafty yourself, consider supporting the efforts of others who are. The website Etsy.com is a fantastic resource for handmade creations of all kinds, and I'll be mentioning Etsy throughout this book. Typing "recycled shopping bag" into Etsy's search box brings up over a thousand different handmade bags made from repurposed t-shirts, jeans, sweaters, skirts, draperies, tablecloths, sheets, and feed bags, as well as knitted or crocheted totes made from homemade plastic bag yarn, or "plarn."

PLASTIC WASTE-BUSTER: Sharon Rowe, founder Eco-Bags Products, Inc.

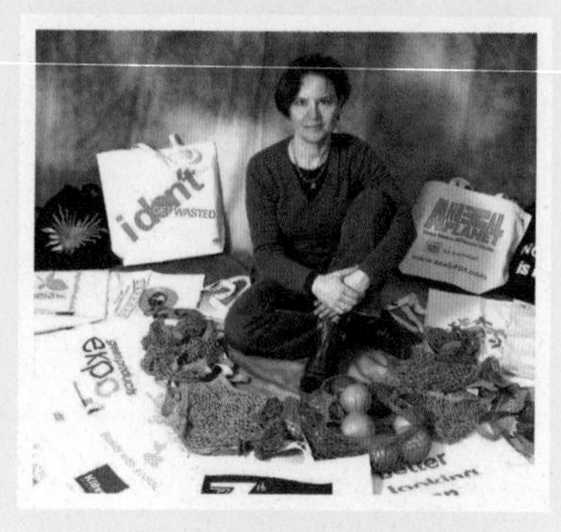

Sharon Rowe was an actor living in Manhattan with her musician husband and new baby when she first woke up to the problem of plastic pollution in the late 1980s. Lugging her groceries home each day, she grew frustrated with the flimsy plastic bags that would fall apart and become instant trash before she even stepped through her door. And looking around her neighborhood at how the plastic trash would pile up in the street, she thought, "I don't want to do this anymore." She told her husband there had to be a way to live without throwing away so much stuff. And then she remembered the net grocery bags she'd seen while travelling around Europe a few years earlier. She tried fruitlessly to find similar bags in the United States and finally asked a friend to bring her back some from Europe. When her New York friends noticed the bags and wanted their own, she knew she was onto a good idea.

But Sharon was an actor, not a business person. She knew nothing about importing and marketing products. But her father had been a retail store owner, and as Sharon told me, "the apple didn't fall far from the tree." Sharon credits her independent personality and ability to improvise with giving her the confidence to start her own business. She started faxing and calling consulates to get information about factories making the string bags. This was back in 1989, before the Internet made global communication so much easier. She got a few responses and wired money to get sample bags made up. Then, she set out on foot, going door to door to show her samples to the businesses on Columbus Avenue. And she started to get accounts. Then, at the big Earth Day festival in 1992, Sharon signed up for a table and had thousands of bags shipped to her NYC apartment. The day of the festival, she sold out of bags in three hours. They were that popular.

From the beginning, it was important to Sharon to make her bags of natural (or recycled) materials grown without toxic chemicals and produced with fair labor standards. She developed strong relationships with her production partners to ensure that the cotton was grown without pesticides and introduced goods with organic cotton as far back as 1994. She's worked with the same factory in India since the beginning of Eco-Bags Products, one which is SA8000 certified by Social Accountability International to follow policies and procedures that protect the basic human rights of workers and which offers health and retirement benefits, as well as vacations and holidays. Additionally, she produces in China and the USA for custom projects, using facilities with stringent social and material certifications in place.

Nowadays, the ECOBAGS brand includes not only the original string bags but a full line of customizable canvas totes, cloth produce bags, lunch sacks, reusable sandwich/snack bags, and bulk bags. Materials include organic cotton, recycled cotton, hemp, recycled bottles, woven grasses, and more. All of the products are made to last a long time. Back when Eco-Bags Products first started, there weren't many reusable bag companies to choose from. But nowadays the market is saturated with both good quality bags and cheaply made ones. According to Sharon, you can make money by selling crappy bags that will fall apart and have to be replaced over and over (which are not much better for the planet than disposable bags), or you can sell just a few bags to a lot of people. She wants to be in the latter camp, growing her customer base without filling up people's closets with junk. She likes that her customers tell her that they've been using the same bags for over fourteen years. "The brands carry our stories," she says. "They show the values of the company."

I asked Sharon what advice she would give to someone else who had an idea for a product that would help us reduce our addiction to plastic but wasn't sure how to get started. She said, "Just start. Make the prototype. Start using it. Start talking to people about it. Talk to people in the industry where you think it might work. See if it's viable. And make sure the market is not already saturated." She also said that a willingness to live simply can help. "Simplifying your own life helps in the beginning before you start making money."

Make Your Own Reusable Cloth T-Shirt Bag

Because of their shape, plastic grocery bags are commonly referred to in the industry as t-shirt bags. Why not make your own t-shirt bag out of an actual shirt? While making the bags is easier with a sewing machine, they can also be made without any sewing at all.

1. Cut the sleeves off an old t-shirt, cutting outside the shoulder seam. Leaving the shoulder seam will make the handles stronger. Or start with an old tank top instead and don't worry about cutting off sleeves.

2. Cut around the neck line on both sides of the shirt. Hemming the edges is not necessary.

3. Turn the shirt inside out and sew a double seam across the bottom. Using a zigzag stitch will make a stronger seam. Alternatively, cut fringes across the bottom and tie the matching front and back fringes together. No sewing necessary.

4. Flip the shirt right side out again. Voila! Your bag is ready to use.

5. Optional: Consider making pockets for your bag out of the cut off sleeves.

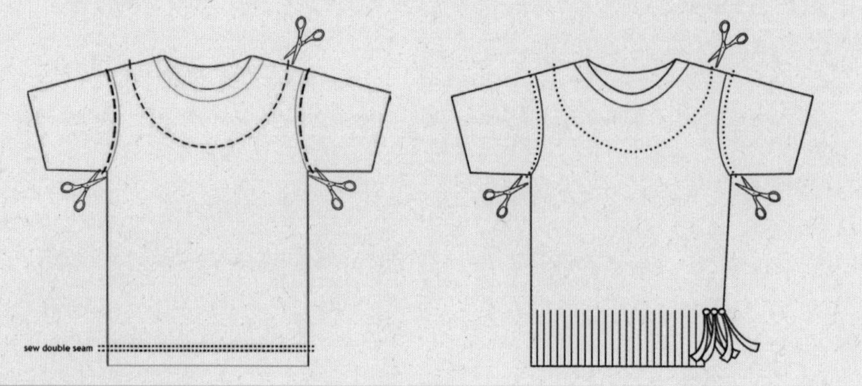

New Bags

Organic Cotton, Recycled Cotton, Hemp, or Burlap Bags If you're going to buy new reusable bags, natural, organic plastic-free materials are a good option. I don't generally advocate purchasing conventional cotton products because cotton is typically grown using a lot of petrochemical pesticides and fertilizers. But organic cotton is a good choice. And recycled cotton uses even fewer resources. Hemp requires much less water and resources to grow than cotton. And all of these fabrics can be tossed into the washer with the rest of your laundry. Here are just a few of my favorite brands:

- **ECOBAGS** Made from organic cotton, recycled cotton, and hemp blends.
- **Project GreenBag** Made of organic cotton in the United States.
- **Rejavanate** Made from recycled burlap coffee sacks by developmentally disabled adults.

Synthetic Bags While the purpose of my project is to avoid buying new plastic, and synthetic polyester bags are indeed a kind of plastic product, I decided early on to make an exception for ChicoBags. ChicoBags are lightweight and many are shaped just like disposable grocery bags, but the fabric is tough and long-lasting. They fold up into their own integrated stuff sack, which can't get lost since it's attached to the bag itself. A drawstring keeps the sack shut tight and compressed smaller than a balled up pair of socks until you're ready to use the bag. I keep a few ChicoBags in my purse at all times, and I have even pulled them out in a department store once or twice to carry new clothing or cookware. (Yes, I've come a long way from that first embarrassing attempt at using my own bags in the grocery store.) Like other cloth bags, ChicoBags can be washed with the rest of your laundry over and over again. Choose ChicoBag's rePETe style bags, which are primarily made from recycled bottles, to avoid purchasing brand new plastic. And when the bags finally wear out (which I'm guessing will take a long time since mine are over seven years old and still going strong) send them back to ChicoBag's Repurposing Program, where they will be made into new products or, if

PLASTIC WASTE-BUSTER: Andy Keller, founder ChicoBag

In early 2004, standing in front of an enormous mountain of trash—what his town of one hundred thousand people discarded in one day—Andy Keller realized he was part of the problem. He had come to the landfill to drop off a truckload of yard waste. Having just been laid off from his job as a software salesperson, he had plenty of time to garden and think about what to do next with his life. He knew that the sales jobs available in pharmaceuticals or text books or software didn't appeal to him, but he wasn't sure what did. What he saw before him that day at the landfill—piles of plastic bags blowing in the wind, pecked by birds, destined to last for a very long time—shocked him into action. Right then, Andy decided to stop using plastic bags. And that decision changed the course of his life.

Making the switch to reusable bags wasn't easy. They were bulky and hard to remember. And if it was hard for him, it must be even harder for people who hadn't seen what he had. What solution could he come up with, for himself and for others, to make quitting plastic bags easier? Andy knew how to sew, having learned on his mom's sewing machine as a kid making Santa hats for his stuffed animals, and he'd always had an entrepreneurial spirit, mowing lawns and fundraising to win a portable radio at school. He'd studied business and marketing in college. So he was confident he could come up with a solution to the reusable bag problem that would not only help the planet but solve his unemployment problem as well. Andy picked up a secondhand sewing machine from the thrift store, sat down at his kitchen table, and designed what would eventually become a ChicoBag—a strong, lightweight full-sized reusable bag you can easily carry in your pocket. In 2008, he introduced bags made from recycled materials in order to reduce the environmental impact of his synthetic bags, and early on he created a take back program so that worn out bags would not end up as more landfill waste.

Andy could have been satisfied just selling his reusable bags, but he is an activist at heart. Wanting to find a way to give people the experience he had at the landfill that day in 2004, he got the kids at a local school to help him collect 500 used plastic bags—the number of bags the average consumer uses in a year, which they connected into a chain that wrapped all the way around the school. He rolled the bag chain up into a huge ball that he would bring with him to events to demonstrate the extent of the plastic bag problem. At one event, he hid under his pile of bags and jumped up with the pile to scare people. That day, the now infamous Bag Monster was born. Afterwards, Andy worked with a local artist and made an actual Bag Monster costume, sewing the bags to a canvas base that could easily be worn. There are now about a hundred Bag Monsters, which groups can borrow for educational purposes. (Email bagmonster@chicobag.com to invite the Bag Monster to your event.) I had the chance to try on the Bag Monster costume at the Green Festival in San Francisco back in 2009, and I have to say, it's pretty cozy—and a lot more comfortable than the plastic sea monster costume I made for myself that year (see chapter 11).

I asked Andy the same question I asked Sharon: what advice would he give to someone who had come up with a creative business idea for reducing plastic waste but didn't know what steps to take? He suggested first finding a mentor. He found his through his local Small Business Development Center, which offers free consulting and help. Then, he said, "Think your idea through; test it out; get advice to see if it can be a business. But above all, take that first step and don't delay. Don't get stuck in analysis paralysis. You don't need a big business plan and investors to get started. You will get on-the-job training." Maybe, as you read this book and understand some of the challenges to plastic-free living that remain, you'll come up with a creative idea or two yourself. I'm rooting for you!

still functional, distributed to low-income families for further use. In ChicoBag's most recent repurposing project, the company worked with victims of domestic violence in Pennsylvania to make woven rugs and drink coasters. In fact, ChicoBag will take back all of your old reusable bags, no matter what brand. ChicoBag is always looking for new repurposing ideas and partners.

> "Get cloth shopping bags that have the handle material sewn into the sides right down to the bottom of the bag. Cheap reusable shopping bags may not save anything in the long run if they don't hold up."
> —Martin Higgins, www.plasticless.com, Malta

Reusable Bags to Avoid

Some reusable bags are not much better than disposable bags. Those cheap nonwoven polypropylene bags grocery stores sell for ninety-nine cents look like fabric, but they are actually plastic. They are not recyclable, not washable, fall apart quickly, and are mostly made in China. In January 2011, *USA Today* reported that nearly a third of the nonwoven polypropylene bags handed out by forty-four major retailers contained high levels of lead.[49] And the bags are so inexpensive that people start to accumulate them the same way they do disposable bags. Avoid the temptation to relieve your conscience by purchasing another ninety-nine-cent bag. Honestly? I'd rather carry stuff out in my hands, which I actually did a while back. Read on . . .

Remembering Our Bags

The biggest obstacle to reusable bag use is remembering to bring them to the store. But it's really just a matter of creating a new habit. Here are a few tips:

Keep Reusable Bags in Your Car. Small bags can be rolled up and stashed in a cup holder where they're easy to see. If you don't drive a car, keep some hanging on your bike or whatever vehicle you use to get around. Replace them immediately after emptying your purchases so they'll be there for you the next time.

Store Reusable Bags Wherever You Keep Your Wallet and Keys. You wouldn't leave the house without those. And once again, replace them immediately after emptying them.

Stash Some Small ChicoBags or Reused Plastic Bags in Your Purse, Pockets, or Wallet, So You'll Have Them Automatically. You never know when you might need a bag. Replace immediately.

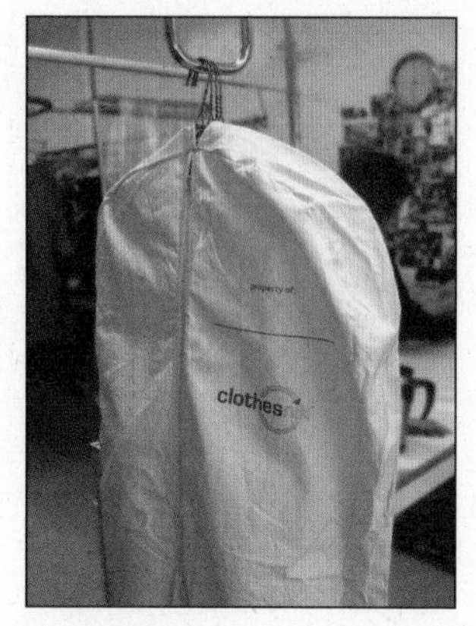

Clothesnik reusable cloth dry cleaner bag

Quick Tip:

Refuse plastic dry cleaner bags. Consider purchasing a Clothesnik reusable canvas dry cleaner bag and ask your cleaner to use it instead. Just bring it with you when you drop off your clothes.

Don't Let Yourself Off the Hook Too Easily. It requires repetition to create a new habit. If you're at the store and have left your bags in the car, take a minute to go back out and get them. It's not a punishment for forgetting but a way to reinforce the new behavior you want to develop. An environmental blogger I met told me that one time, after forgetting to bring her bags to the store with her, she paid for her groceries and then loaded them back into her shopping cart to wheel out to her car, put them in the trunk without bags, drove home, then brought her reusable bags out to the car and filled them up to carry her purchases into the house. That's dedication.

Use Your Hands . . . or Shirt! Early on in this project, Michael and I found ourselves without bags while shopping for groceries on vacation. We were in a new environment and had completely forgotten about bringing our bags. Pushing a cart full of fresh produce to take back to our hotel, we searched every section of the store for some kind of paper bag to use instead of plastic, but there were none. Finally, we ended up holding our t-shirts out in front of us to transport our

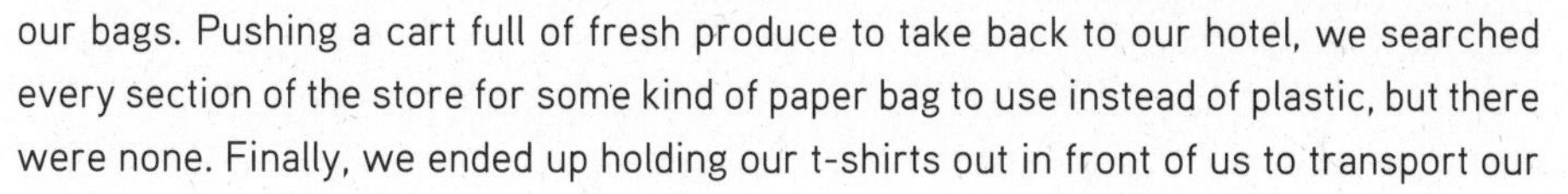

apples, oranges, avocados, and melons. ***Melons in my shirt.*** Embarrassing? Yes. Extreme? Depends on who you ask. But also fun. We came home with a funny story to tell. And do you think we ever forgot our reusable bags again? Nope.

Take It Slow. For those who don't see themselves going the extreme route, Andy Keller recommends just getting in the habit of bringing one reusable bag with you to the store every time and replacing one disposable bag with a reusable one. Once that's a habit, add more.

Forget Guilt. If and when you do end up with a disposable bag, don't beat yourself up over it. Just commit to remembering your bags the next time and figure out what strategies you can use to help yourself remember.

Wash Those Cloth Bags!

Reusable bags get dirty, just like our clothes. We wash our clothes. It's just common sense to wash our cloth bags, right? And yet a 2010 study of reusable shopping bags collected from consumers entering grocery stores in three major cities found that all of the bags contained high levels of bacteria. Most of the bags tested, it turns out, were those cheap nonwoven polypropylene bags I mentioned above. And most customers admitted to never washing their bags. But the same study also found that bacteria levels in bags could be reduced 99.9 percent by hand or machine washing. And the potential for cross contamination is significantly reduced by keeping raw meat products separate from other foods.[50] The moral? Choose good quality washable cloth bags, and wash them. Adding a few cloth bags to your laundry load is not going to increase your water or energy usage by much. But it will ensure your safety and that of your family.

Recycling Plastic Bags

So what do we do with the plastic bags we already have after they are worn out and can't be reused as bags anymore? Recycling options do exist. But recycling plastic bags is problematic. First of all, most municipalities don't allow them in curbside programs because they clog up the sorting machines when mixed with other recyclables, a phenomenon I witnessed firsthand during my visit to a California recycling center. (I'll tell you more about that adventure in chapter 4.) So to recycle them, you have to bring them back to the store or recycling center. In 2007, the State of California mandated that all supermarkets and large retail stores with a pharmacy provide plastic bag drop off bins to collect bags for recycling. But the program is not working. In 2009, California stores collected back only 3 percent of the bags they handed out.[51] Where are the rest? You can't convince me that 97 percent of the 53,000 tons of plastic bags handed out in California that year were used to line trash cans and pick up poop. My guess is that a lot of those bags are blowing in the wind.

So what happens to the bags that we do bring back for recycling? Like most of our plastic recycling these days, the majority of plastic bags and other film are exported to China. (I'll discuss issues with overseas recycling in chapter 4.) Of the bags that are recycled domestically, the bulk are sold to a company called Trex, which manufactures composite lumber out of recycled plastic film, scrap wood, and sawdust. But recycling plastics bags into plastic lumber is actually downcycling because it doesn't reduce the demand for brand new plastic bags and because Trex lumber cannot be further recycled. Once Trex wears out, it ends up in the landfill. And while there is a growing demand for recycled plastic bags to be incorporated into garden products, crates, buckets, pallets, and piping, those markets are still very small.

Can't plastic bags be recycled into new plastic bags? While it's true that a very small percentage of plastic bags do get incorporated into new bags, there is some dispute as to whether plastic bags can actually be made from 100 percent recycled content.[52] Most recycled plastic bags contain a combination of both recycled material and virgin plastic resin. And why do we need to continue producing more plastic bags in the first place? Whether made from recycled material or not, plastic bags create havoc when let loose in the environment. What's more, no matter what products we recycle our bags into, the fact is that plastic can only be recycled a finite number of times before it loses tensile strength and must finally be retired. While the same is true for paper, whose fibers get shorter the more they are recycled, paper will biodegrade at the end of its life. Plastic will not. Those huge molecules will still be around long after we're gone.

So is it worth it to bring our plastic bags back to the store for recycling? Yes. But not until we've gotten as much use out of them as we can. Let's first reduce the number of new bags we consume, reuse them as much as possible, and only then bring them back for recycling. Recycling is not a solution to the plastic bag problem. It simply keeps them out of landfills and the environment for a little longer.

Plastic Bag Actions: Bans and Fees

After several months of bringing my own reusable bags to the store with me, I had developed a pretty strong habit. I remembered to speak up every time I went through the checkout line and to have my bags ready when it was my turn. It felt good to know I wasn't personally contributing plastic bag pollution to the environment. But looking around me at all the customers still leaving the store with their hands full of disposable plastic, I wondered if my actions were making a difference in the bigger picture. Sure, more and more people were starting to wake up and change their habits. But was it enough? It was clear we needed action on a grander scale. And that's how I found myself one sunny January morning standing with my friend Nancy on the steps of Oakland City Hall with a big sign in support of my city's initiative to ban plastic bags altogether. I hadn't planned to become an activist when I started this plastic-free project, but after doing what I could personally and realizing it just wasn't enough, I felt compelled to take the next step.

Recognizing the harm to the environment and costs to cities to clean up and dispose of plastic bags, many municipalities have begun assessing fees or instituting outright bans on plastic bags. In 2002, the Republic of Ireland became the first country to institute a plastic bag fee to reduce its consumption of 1.2 billion plastic bags per year. The fee was a huge success. Disposable plastic bag use dropped 90 percent.[53] Many other countries and municipalities followed suit. In January 2010, Washington D.C. introduced a five-cent fee for plastic bags in order to fund a program to help clean up the plastic-polluted Anacostia River. The program turned out not to be the revenue-generator the city had expected, as plastic bag consumption dropped by 50 percent that year.[54] But really, what better way to clean up a river than reduce the amount of trash generated in the first place?

San Francisco, on the other hand, was stymied in its attempt to assess fees for plastic bags. Anticipating SF's impending plastic bag fee, the plastic bag industry lobbied for and won a statewide initiative (AB 2449) in 2006 promoting plastic bag recycling but prohibiting cities from charging fees for plastic bags. In response, San Francisco enacted an outright ban instead. And many other California cities have attempted to follow suit. But while San Francisco has been able to maintain its plastic bag ban, most others have been challenged in court by the plastic bag industry, which asserts that since banning plastic bags will drive up consumption of paper bags (which have their own negative impacts), California law dictates that cities must complete Environmental Impact Reports before implementing bans of plastic bags. EIR's are cost prohibitive for many cities, including my city of Oakland. Shortly after passing our plastic bag ban, Oakland was forced to shelve the new law as a result of the industry's legal challenge.

Those of us who had felt so excited the day our plastic bag ban passed were disappointed, of course. But since then, local campaigns, largely spearheaded by California's Clean Seas Coalition, have resulted in one city after another passing bans on plastic bags that have held up in court. And on September 30, 2014, California Governor Jerry Brown signed into law the very first state-wide plastic bag ban in the nation. The new law not only bans plastic bags but also requires merchants to charge a fee for recycled paper bags. The goal is to encourage shoppers to bring their own reusable bags rather than relying on free disposable ones. But whether we ban plastic bags or tax them, we need to get our decision-makers involved if we want to eliminate plastic bag pollution once and for all.

Does your city ban or tax disposable bags? The ChicoBag website has a comprehensive world map of plastic bag initiatives. Click "Track the Movement" at the bottom of the page to check and see if your city, state, or country is listed. If not, why not consider getting involved in a campaign? There are plenty of activist resources listed at the end of this chapter, and on page 76 is an inspiring story of what a few local people from my neck of the woods did to get plastic bags out of their town.

Taking Action in Your Community

Would you like to see plastic bags banned or taxed where you live? Check out this list of helpful resources.

Plastic Bag Initiatives and Actions These sites can help.

- **ChicoBag Take Action Page** ChicoBag has compiled a great list of resources for plastic bag action. Visit to learn about plastic bag legislation in your area, invite the Bag Monster to an event, understand the facts about plastic bags, and learn 10 things you can do to start a reusable bag habit. Contact advocacy@chicobag.com for more information about starting a campaign.
- **Plastic Pollution Coalition's Plastic-Free Towns Page** The PPC also has a list of plastic bag resources that can help you learn the facts, build the movement, connect with others working on plastic bag initiatives, and contact legislators. Visit to start creating your own Plastic Free Town.
- **Plastic Bag Laws** This site is a great resource for tracking plastic bag legislation.
- **Green Sangha's Rethinking Plastics Campaign** Visit to learn more about Green Sangha's Rethinking Plastics Campaign and plans for grocery stores to phase out single-use plastic bags.

Film, Video, and Music About the Plastic Bag Problem For the short videos, I have listed the URL from the sponsoring organization's website where possible. But you can also find these videos by typing their titles into the YouTube.com search box.

- ***Bag It* Documentary Film.** Presented in a compelling yet humorous way, this award-winning film not only explains the problems with plastic bags but also outlines the steps the plastic bag industry has taken to defeat plastic bag initiatives. Once it motivates you to get plastic bags out of your life, it'll make you rethink all other kinds of plastic as well. *Bag It* is a great community organizing tool. Visit www.bagitmovie.com to learn how to find a local screening or host a screening yourself. You can buy the film on DVD or rent it from Netflix. And the *Bag It* website has educational materials for teachers to accompany the film. Look for cameo appearances by me and a few of the other characters profiled in this book. Click the BagIt Town link for activist tools to help your town go plastic bag-free.

- **Plastic State of Mind Parody Music Video** Sponsored by Green Sangha, this catchy and informative video about plastic bags is based on a wildly popular hip hop song. Available on YouTube and via the Green Sangha website.
- **The Majestic Plastic Bag, a Mockumentary** Narrated by Jeremy Irons, this video is a spoof of David Attenborough-style nature documentaries. It begins, "The open plains of the asphalt jungle, home to many creatures great and small, and the pupping ground for one of the most clever and illustrious creatures, the plastic bag." It goes on to detail the realities of plastic bag pollution in a brilliantly inventive way. Available on YouTube.
- **The Bay vs. the Bag** Created by Free Range Studios, the same company that produced Annie Leonard's viral Story of Stuff video, this stop action animation uses a giant plastic bag wave to illustrate the immensity of the plastic bag problem. View the video on SaveSFBay.org's "The Bay vs. The Bag" page or on YouTube.
- **Plastic Bag, by Ramin Bahrani** Voiced by the inimitable Werner Herzog, this short film traces the epic, existential journey of a plastic bag searching for its lost maker, the woman who brought it home from the store and then discarded it. The ending is devastating. View the video on the Future States website or on YouTube.
- **Canvas Bags, by Tim Minchin** This hilarious music video begins with comedian Tim Minchin as an awkward accordion-playing geek performing for his web cam and morphs into a full-blown, wildly addictive anti-plastic bag

PLASTIC-FREE HEROES:
Renee Goddard & Andy Peri

Across the San Francisco Bay in a little town called Fairfax, Renee Goddard was shocked into action after seeing Al Gore's film *An Inconvenient Truth*. She didn't consider herself an activist and had no prior experience in community organizing. But having been a white water river guide, she knew how to shout out commands to motivate people to action. She put up flyers around town inviting residents to meet up at a park to discuss ways to make their town more environmentally friendly. Around sixty people showed up to the first meeting on July 18, 2006, and compiled a huge list of ideas: support farmers markets, designate half the road for bikes once a week, hang laundry to dry, go solar, consume less, and many more. They all agreed that they wanted to support each other in reducing plastic use.

The group dubbed themselves The Inconvenient Group because they would meet once a week at 5 PM even though most of the members worked and had kids. During one of their meetings, they invited the Bay Area group Green Sangha to come and give its Rethinking Plastics presentation, which highlights the history and problems of plastic. Larry Bragman, then a member of Fairfax's town council (who would later become the mayor of Fairfax) attended that presentation, and shortly thereafter he, Green Sangha, and the Inconvenient Group began to talk about banning plastic bags in Fairfax. Bragman wrote up the legislation, modeling it after the San Francisco law, and then worked with the community to support it.

I sat down with Andy Peri from Green Sangha to hear his side of the Fairfax plastic bag ban story. Andy explained that he wanted to make sure they'd have support from the business community before passing this legislation to avoid backlash. So one

Saturday afternoon, he made up a half-page flyer, got on his bike, and chose one side of the street in the business district to canvass. At each store, he asked the owner or manager how they'd feel about a ban on plastic bags. He was particularly concerned about the local lumber and hardware store, Fairfax Lumber, and other stores he knew used plastic bags routinely. But store after store gave the idea of a ban their full support. In fact the Fairfax Lumber store manager told Andy that as a scuba diver, he saw plastic bags on the bottom of the ocean all the time. He said he would never support having plastic bags at the checkout stand. While some business owners had concern with timing and other details, not one business on that street opposed the measure.

The Fairfax City Council passed Ordinance 722 banning plastic checkout bags on July 11, 2007, and shortly thereafter a fax came from the plastics industry threatening a lawsuit. To avoid litigation, the town council amended the ordinance to make it voluntary.[55] Andy says he wasn't surprised by the lawsuit, but had kind of hoped that the plastic bag industry wouldn't care so much about a little town like Fairfax. I asked how the group members felt after getting word that the town was being sued. He sighed, "There was definitely a quality of resentment that a community that had a very consistent environmental ethic and goals, that was very well-informed, would be challenged by a giant corporation that only had its self-interest in mind." That's the nice, tactful, Zen way of saying they were pissed off.

But the story doesn't end there. Because the way California law is written, if the initiative were presented to and passed by over 50 percent of Fairfax voters, it would be immune from industry lawsuits. So Renee, Andy, and the gang regrouped and set out to gather the necessary signatures to qualify for the 2008 ballot. They set up an information table at the Fairfax farmers market and at the Good Earth Natural Foods store for passing out information and collecting signatures. Renee got together with other moms and sewed 100 reusable bags to give away. Friends went door to door with flyers and clipboards. They invited the ChicoBag Bag Monster to come and pay a visit to the town. Finally, in November 2008, Fairfax passed its plastic bag ban with 79 percent of the vote.[56]

Renee says, "It was a celebration of local democracy."

anthem, complete with marching band and wind machine. The night I first saw it, I must have hit the Replay button twenty times or more. Available on YouTube, of course.

- **Turtle Ate a Jelly** Adorable and catchy kids' song about keeping plastic bags out of the sea from the Banana Slugs on the album *Only One Ocean*.

Action Items Checklist:

(Choose the steps that feel right to you. Then, as an experiment, challenge yourself to do one thing that feels a little more difficult. Only you know what that one thing is.)

- ☐ Collect a stash of reusable bags.
- ☐ Make a plan to carry/remember them.
- ☐ Practice bringing bags with you when you leave the house.
- ☐ Talk to your friends and family about why you have decided to stop accepting disposable plastic bags. Explain, but don't nag! Seriously.
- ☐ Bring an extra bag to offer someone in the checkout line who doesn't have one.
- ☐ Speak to a store manager about eliminating plastic bags.
- ☐ Watch the film *Bag It* alone or with friends.
- ☐ Organize a screening of the film *Bag It* through a civic group, church, PTA, or other organization.
- ☐ Organize a reusable bag making party with or without kids.
- ☐ Write a letter to the editor of a local paper explaining the problems with plastic bags.
- ☐ Find out what plastic bag initiatives are happening in your area.
- ☐ Write to a legislator about banning or taxing plastic bags.
- ☐ Join a group campaigning against plastic bags.
- ☐ Start your own plastic bag reduction campaign.

Chapter 3: Plastic Beverage Bottles (Dealing with a Drinking Problem)

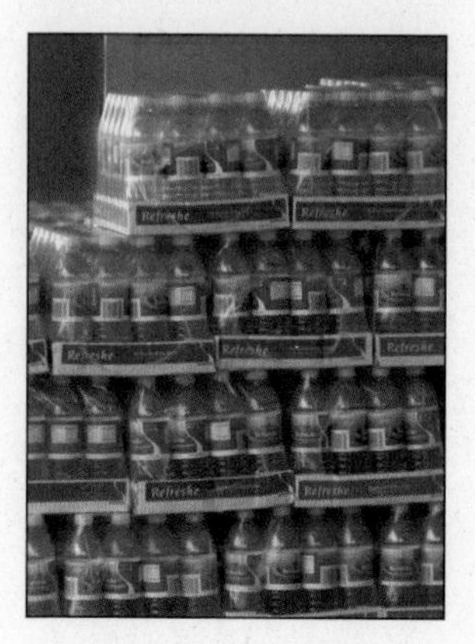

After giving up disposable plastic shopping bags, I decided that my next step would be to quit buying bottled water. As a distance runner, I already had a stash of reusable sports bottles at home. And living in the Bay Area, I had a fresh clean supply of drinking water running from my own tap, nearly free! Why was I spending $1.50 for a new bottle of Dasani or Aquafina every time I went to the gym when with a little planning ahead, I could bring water from home? Or fill up a bottle from the free water fountain on the workout floor? I didn't realize then how important this step would be. In the past few years, I've learned that the environmental impact from bottled water is about more than just the plastic bottle. But back then, it just seemed like an easy way to reduce my consumption of new plastic. The biggest challenge was remembering to bring my bottle.

I kept a mental list of the gear I needed to stash in my backpack for the gym: athletic shoes, socks, shorts, towel, lock, and key. Now I just had to add a reusable bottle. The first few times, I remembered it. And then one evening, I forgot. Standing half-dressed in the locker room, rummaging frantically through my backpack, I realized I hadn't packed a bottle. Panic set in. Holy crap. How could I spend thirty to sixty minutes on a cardio machine without water? I was sure I would shrivel up and die. Everyone had water with them. Everyone. There were bottle holders attached to the treadmills and elliptical trainers and stair climbers, which meant that even the equipment manufacturers understood the importance of constant hydration! What should I do? Give in once again to the vending machine? Leave without working out? Buy a new (plastic) sports bottle from the front desk? Or do the nearly unthinkable . . . work out without water?

And then I remembered Greenbelt Lake. Back when I was in college in the mid-1980s, my dad and I took up running. He would meet me at my apartment first thing in the morning, I'd throw on my running shoes, and we'd head down the street to join the other fitness buffs doing laps around the man-made lake in my suburban Maryland neighborhood. We each drank a glass of water before leaving the house and again when we got back home. There might have been a water fountain in the park for rehydrating between laps (except I don't think we called it "hydrating" back then), but we certainly didn't buy bottled water. The only bottled water brands we'd heard of were Perrier, which was for rich people, and Evian, which we joked was "naïve spelled backwards." Why would anyone pay money for expensive water in a bottle? The point is that we survived our runs without carrying water. The memory calmed me down, and that night at the gym, I made a plan. I would drink from the water fountain before my workout, and if I needed to, I could actually get off the machine and walk to the fountain to drink some more. The plan worked. I completed my routine without a bottle of water by my side, and I didn't die.

I'll admit that going back and forth to the water fountain was not exactly convenient. It's much nicer to have my own bottle by my side, and these days I rarely forget to bring my reusable bottle or travel mug with me when I go out. But realizing that I could survive without the instant gratification that disposable plastic provides was actually pretty empowering. The bottled water industry spends over $150 million per year on advertising in the United States to convince us we can't be healthy or satisfied without its product. I had bought into the bottled water myth. And that night at the gym, I broke free.

The Truth about Bottled Water

Is bottled water tastier, healthier, and more convenient than tap water? The bottled water industry would like you to think so. Using scare tactics to convince us that our tap water isn't safe, labeling its bottles with images of arctic mountain springs, and paying big bucks to celebrities to tout the health and beauty benefits of their brands, bottled water companies have manufactured demand for a product that barely existed thirty years ago. The Container Recycling Institute estimates that Americans now consume about 42.6 billion single-serving (1 liter or less) plastic water bottles each year.[57] We've bought into the industry's marketing claims. But how does the truth about bottled water compare to the hype? Here are a few facts:

Myth #1: Bottled Water Is Safer Than Tap Water

Fact: Bottled water is less regulated than the water that runs from our kitchen faucets and public drinking fountains. In the United States, bottled water and municipal tap water are actually regulated by two different government agencies. The Environmental Protection Agency (EPA) sets the standards for tap water under the Safe Drinking Water Act, which requires public water systems to be tested regularly by a government-certified laboratory, that violations be reported to the public within a specified period of time, and that water quality results be made available annually. Those of us whose tap water comes from a municipal system can obtain a copy of our Consumer Confidence Report (a.k.a. Annual Water Quality Report) to find out where our water comes from, what contaminants, if any, are in our water, and how their levels compare to EPA safety standards. Many utility districts provide these reports online. On the other hand, bottled water is regulated by the Food and Drug Administration (FDA) under the Federal Food, Drug, and Cosmetic Act (FD&C) and while most of the FDA's standards for contaminants in water mirror those of the EPA, the FDA does not have the same authority as the EPA to regulate water bottling companies. For example, water bottlers are allowed to do their own tests rather than being required to use certified labs and they are not required to provide their test results to the public. What's more, the FDA can't even require bottlers to report test results if high levels of contaminants are found![58]

So what does this mean in terms of the safety of bottled water compared to tap water? A glance at your municipal water quality report might look scary. All kinds of pollutants can get into our tap water from agricultural runoff, factory emissions, and the byproducts of purification itself. But does that mean bottled water is safer? The Environmental Working Group wanted to find out. So in 2008, the organization tested ten popular brands of bottled water and found a shocking thirty-eight different pollutants, those found in municipal tap water as well as a whole category of chemicals routinely used in plastic production. Pollutants found in bottled water included byproducts of water treatment (trihalomethanes, haloacetic acids, fluoride), fertilizers (nitrate, ammonia), drugs (acetaminophen, caffeine), plastic production chemicals (acetaldehyde, hexane, toluene, and nineteen others), as well as bacterial contamination, arsenic, and radioactive pollutants. All of the brands tested contained some of these pollutants, and two store brands—Wal-Mart's Sam's Choice and Giant's Acadia brand—contained levels that exceeded California's safety standards.[59] There is no guarantee that bottled water is safer than what comes out of your tap, and unless bottled water companies voluntarily post their test results (to be fair, some of them do), there's no way for consumers to know for sure what they're drinking.

There are areas of the world where the tap water is too polluted to drink. And there are emergency situations in which bottled water might be necessary. But for most of us in the United States, why are we paying for a product that costs at least a thousand times more than what runs from our taps when we aren't even guaranteed that it's safer? There's a better solution. Purchasing a good water filter can help remove contaminants in our tap water for less than the cost of all of those plastic bottles. (You'll find information on testing your tap water and choosing a water filter later in this chapter.)

Myth #2: Bottled Water Tastes Better Than Tap Water

Fact: Most people can't taste the difference between bottled water and tap water. In 2003, the irreverent team of debunking magicians, Penn & Teller, set out to expose the myths of bottled water on their Showtime television series, *Bullshit!* First, they did an unscientific survey on the streets of New York City, asking passersby to blind taste two different waters. One came from a nearby tap and the other was a "relatively expensive store-bought bottled water." The result? According to Penn, 75 percent of the people who took the test were surprised to find they had chosen New York City tap water over the bottled water. Spurred on by the success of that little experiment, the guys went further to demonstrate how bottled water marketing itself impacts not only consumers' expectations but their actual perceptions of how water tastes. In a hilarious segment, Penn & Teller took over a trendy California restaurant, installing their own made up "water steward," a fake employee who visited each table and presented diners with a menu of exotic and refreshing-sounding waters from around the world. What patrons didn't know was that all of the bottles had been filled from the same hose out back behind the restaurant. Based on the images on the made-up labels, the prices of the waters, and the stories told by the water steward, diners actually believed they could taste differences among the various waters and were shocked to finally learn that all of the waters were nothing more than Los Angeles tap water.[60]

So can you believe what you see on TV? Reality shows can be edited to prove pretty much anything. But I personally witnessed similar results back in 2007 when I joined up with Corporate Accountability International to help promote its Think Outside the Bottle Campaign. Setting up taste tests on college campuses, at farmers markets, and at press events around the country, volunteers blindfolded participants and asked them to taste four different waters: two from municipal sources and two different brands of bottled water. I participated in one such event on the steps of Oakland's City Hall in October of that year and watched as one by one, taste testers guessed wrong or preferred the city's tap water over the bottled waters. Without the visual cues provided by millions of marketing dollars, we can rely on our taste buds to tell the truth. And often, the truth is that bottled water doesn't taste better. So why pay for it?

Of course, some tap water does taste pretty crappy. I was disappointed by the stuff that came out of the kitchen faucet during a trip to Orlando, Florida, last year. (No offense, Orlando. We like what we're used to, right?) But the solution to yucky tasting water isn't bottled water. The same home water filter systems that can improve the purity of tap water can also improve its taste, for way less cost and without all the plastic bottle waste.

Myth #3: Bottled Water Is More Convenient Than Tap Water

Fact: The inconveniences created by our dependence on bottled water outweigh the few conveniences it provides. While it may seem convenient to be able to quench our thirst from the nearest vending machine or minimart pretty much whenever we want, consider the inconveniences associated with bottled water. First of all, bottled water is expensive! It costs over a thousand times more than tap water, and when purchased in single-size bottles, costs even more than gasoline. A quick check on Safeway's website tonight showed twenty-ounce bottles of Aquafina and Dasani selling for $1.79/bottle.[61]

That's $11.45/gallon! The U.S. average gas price tonight is $3.62/gallon.[62] But it makes sense, when you think about it, that bottled water would cost more than gas. According to Peter Gleick, founder of the Pacific Institute and author of the book ***Bottled and Sold: The Story Behind Our Obsession with Bottled Water***, it took approximately 17 million barrels of oil equivalent to produce the plastic for bottled water consumed by Americans in 2006—enough energy to fuel more than 1 million cars and light trucks for a year—and created more than 2.5 million tons of CO_2. Additional energy was required to fill those bottles, ship them, cool them, and finally collect them again for recycling or landfilling. Bottled water is a waste of money and resources. That's inconvenient.

And bottled water is inconvenient for communities where the water is extracted. According to the U.S. Geological Survey, water extraction can alter local groundwater levels. It can lower the local water table shared by nearby well users and affect natural resources dependent on groundwater flowing, such as fish and other wildlife populations. Basically, extracting water from one area and shipping it to another disrupts the natural water cycle that we all depend on.[63] And the group Food & Water Watch reports that the jobs the bottled water industry brings to a community in exchange for its water are few, low-paying, and dangerous.[64]

Bottled water is especially inconvenient for the people who live and work where the petrochemicals for the plastic bottles are produced. PET, the kind of plastic used for most bottled beverages, is manufactured using a toxic soup of chemicals, one of which is paraxylene, a derivative of benzene, a highly carcinogenic chemical extracted from crude oil through a refining process at oil and petrochemical refineries. The documentary film ***Tapped*** interviews residents near one of the largest of these facilities, Flint Hills in Corpus Christi, Texas, where levels of cancer and birth defects are 84 percent higher than the state average. Isn't it ironic, and inconvenient, that bottled water, which is touted by the industry as being purer and healthier than tap water, is contained in plastic whose manufacture contributes to pollution of our water resources in the first place?

And bottled water is inconvenient for communities that have to deal with mountains of non-biodegradable plastic waste after the contents have been consumed. The beverage industry maintains that curbside recycling programs are the solution to plastic bottle waste. But unfortunately, as with plastic bag recycling, those programs don't go far enough. In 2013, only 31 percent of all the PET bottles on U.S. store shelves were collected back for recycling.[65] One reason for this low return rate is that bottled beverages are most often consumed away from home, away from curbside recycle bins. Those bottles generally end up as waste in landfills and incinerators, or worse, escape into the environment. And who has to pay to clean up all that waste? The taxpayers, that's who.

Myth #4: Bottled Water Doesn't Compete with Tap Water

Plastic water bottles litter the beach.

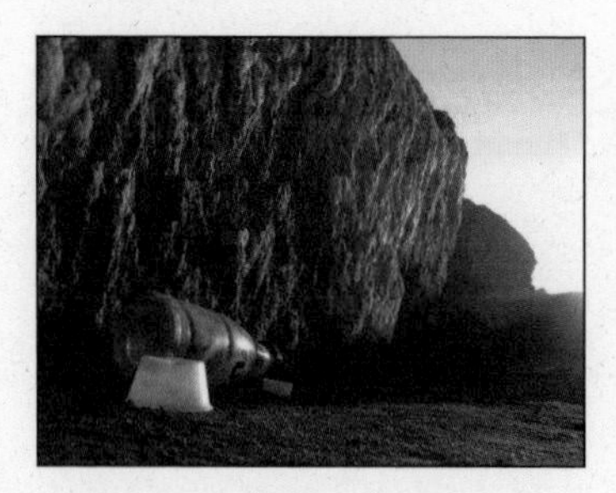

Fact: Reliance on bottled water companies undermines our public water infrastructure. The International Bottled Water Association denies that bottled water competes with tap water.[66] But the truth is when we divert money to bottled water corporations such as Coca-Cola, Pepsi, and Nestle, we spend less on the public water infrastructures that ensure all citizens have access to cheap, clean drinking water. Safe drinking water is a right that shouldn't be reserved for those who can afford to buy bottled water. The organization Food & Water Watch is asking congress to establish a clean water trust fund to upgrade our public water systems. And in a 2010 survey, the organization found that 63 percent of U.S voters favored funding the program with a one-cent per ounce tax on manufacturers of bottled water. Since many brands of bottled water actually come from municipal sources (despite the pictures on the label), I think such a tax would be a great way to transfer some of those profits back to communities where the water came from to begin with!

Back in 2007, I decided to stop drinking bottled water as a way to avoid plastic bottle waste. I had no idea that the impacts of water bottling go way beyond the plastic bottle. Since then, companies have tried to come up with solutions to reduce plastic bottle waste: making bottles with less plastic, making bottles out of plants, bottling water in glass or paperboard cartons, and so on. But none of these solutions address the larger problems of water privatization and the massive amounts of energy used to extract and ship water. The global privatization of water is a huge issue. I've included many resources at the end of this chapter for more information on water issues and ways to get involved. And I've included stories about some inspiring folks working to make tap water convenient and desirable again. For now, let's start with ourselves and the practical steps we can take right now.

"Bottled water companies are taking what's ours to begin with, packaging it in a lethal way, overcharging us for it, and asking us to pay for the recycling. All they're really doing is selling us this convenience, which is actually highly inconvenient. What's convenient about any of the health problems linked with using plastic water bottles?" —Jackson Browne, musician

Choosing a Reusable Beverage Container

Just as with reusable bags, reusable bottles are not created equal. When I first started my plastic-free project, I figured I would just use what I already had: a stash of plastic sports bottles. But the more I read about chemicals leaching from plastic, the less those plastic sports bottles appealed to me. I switched to a stainless steel water bottle pretty early on. Now, I prefer the versatility of a stainless steel travel mug, which can be used for any kind of beverage, from water to coffee or even beer in a pinch! But we're all different and have different habits and preferences. So here, then, is a list of the pros and cons of various kinds of reusable beverage containers. The key is to choose one, use it, and don't lose it!

Glass Mason Jar When No Impact Man Colin Beavan did his one-year experiment, he chose to carry a reused glass mason jar for water, coffee, and other beverages. I asked him once how he drank coffee from it without burning his fingers, and he just teased me about my delicate digits. So I tried carrying a jar for a while myself, and I even made a little cozy for it out of an old sweater to protect my sensitive fingers. I figured, as I've said before and will repeat throughout this book, that the greenest choice was something I already had rather than buying something new. But what I discovered is that there is a spectrum between personal convenience and impact on the planet, and since few of us are saints, our choices will fall somewhere along the spectrum, depending on our lifestyles and how far we're willing to go. For me, carrying a mason jar was inconvenient because (1) I couldn't drink out of it while on the go without spilling, (2) the jar wouldn't fit in the cup holder on my bike, (3) it didn't have a handle, and (4) the jar was breakable. Still, I carried it with me and used it regularly until the day I was refused entrance to a theme park because of a No Glass policy. My little jar was confiscated, and I switched to stainless steel.

Since the first edition of this book was released, I've discovered a plastic-free solution to the first problem: EcoJarz offers reusable stainless steel lids to convert canning jars into to-go mugs. A thin silicone ring around the inside keeps the lid leak-proof but doesn't come into contact with your beverage. The opening in the lid will also fit a stainless steel drinking straw. EcoJarz lids come in two sizes: wide-mouthed or small-mouthed. And by "mouth," I mean the size of the jar opening. But you knew that, right?

Glass Water Bottles Glass is nice because it's inert, doesn't leach chemicals, won't affect the taste of your water, and you can see through it. But it's also breakable, and, as I discovered, is not allowed at all venues. The main difference between a glass water bottle and a mason jar? The water bottle is generally easier to carry and drink from. Here are a few brands of glass water bottles.

- **Love Bottles** are plastic-free narrow-necked glass water bottles made partially of recycled glass and have a ceramic swing top lid with a silicone washer that creates a water-tight seal.
- **Faucet Face** glass water bottles are sized and shaped similarly to disposable plastic water bottles. They come with small plastic twist-off caps.
- **Life Factory** wide-mouthed glass water bottles are covered with a silicone sleeve to protect them from breakage and come with a #5 plastic (polypropylene) twist-off cap with carry loop. Life Factory also offers #5 plastic sippy lids and silicone baby nipples.
- **Bkr** bottles have a wide body and a narrow neck and come with a silicone protective sleeve and #5 polypropylene twist-off cap with carry loop.

Stainless Steel Beverage Containers The obvious advantage of stainless steel over glass is that it won't break. That said, stainless steel can dent, and if treated roughly, can bend out of shape. Other disadvantages are that you can't see what's inside the bottle as easily as you can with glass, and some people feel that stainless steel imparts a metallic taste to liquids, though I personally have not noticed it. While individuals with nickel allergies may need to avoid stainless steel, it's generally considered to be a very safe material for holding foods and beverages. Here is information about a few different brands:

- **Klean Kanteen** designed the original stainless steel water bottle in 2004 as an alternative to plastic and aluminum. Klean Kanteen bottles come in various sizes, from child to adult, painted or unpainted, and with different kinds of caps, including plastic sport caps, plastic sippy spouts, twist-off stainless/plastic hybrid caps (which prevent the contents of the bottle from coming into contact with plastic), and caps with and without carry loops. Klean Kanteen also offers insulated bottles for hot beverages. In 2010, Klean Kanteen released Reflect, the first completely plastic-free water bottle. The logo on the Reflect bottle is etched instead of painted, and the cap is made from bamboo and stainless steel with a stainless steel carry loop and an inner silicone gasket.

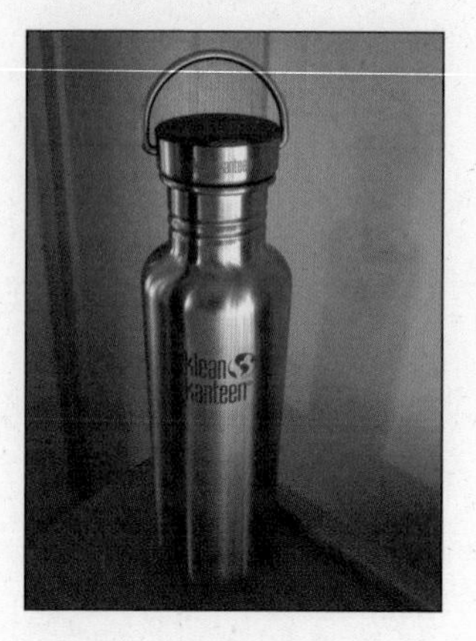

Klean Kanteen Relect plastic-free water bottle

"If you're worried about losing a nice stainless steel or glass water bottle, get one with a loop or ring on the top and attach a pet tag with your name and phone number. It is a fun way to personalize your bottle and the tag will last longer than 'permanent' marker."
—Amy Zehring, San Francisco, California

- **Pura Stainless** bottles look an awful lot like Klean Kanteens. The difference is in the types of caps offered. Pura provides the same stainless/plastic hybrid cap as Klean Kanteen, but unlike Klean Kanteen, the company does not offer any plastic sports caps, plastic sippy spouts, or any other kind of top that allows your lips or your beverage to come into direct contact with plastic. Believing that people should not drink from plastic at all, the company instead offers medical grade silicone sippy spouts and baby nipples.
- **Stainless Steel Mugs** are a more versatile alternative to water bottles since they can hold both hot and cold beverages. There are a whole slew of companies manufacturing travel mugs... too many to name. So I'll just tell you about the one I use. It's the 16 oz. Senja travel mug from Aladdin, which is made from double-walled stainless steel. (Don't be fooled! Some travel mugs are stainless steel on the outside but plastic on the inside.) It fits well in cup holders, and it has a comfortable handle. The travel top is made from #5 polypropylene plastic, which is the main drawback, but because of the way the mug is constructed, I have a choice: I can drink from the opening in the lid while walking, or I can easily unscrew the lid and drink right from the smooth rounded metal edges of the mug itself. If I could find a completely plastic-free travel mug with a handle and with a sip hole for

walking, I'd buy stock in the company, or at least promote it to all my friends. But for now, this is my choice.

Aluminum Water Bottles Aluminum bottles are lightweight, and up until Klean Kanteen came along in 2004, they were the primary alternative to plastic water bottles. Invented by the 100-year-old Swiss company SIGG, aluminum bottles were very popular until concerns about their inside liner began to surface. Unlike stainless steel, aluminum reacts with foods and beverages, so aluminum cans and food containers are lined with a plastic coating to prevent corrosion, a coating that usually contains BPA. At first SIGG denied that there was BPA in its bottle linings. But finally, in August 2009, after changing the formula, SIGG admitted that the lining of its previous bottles had contained BPA after all. SIGG now lines its bottles with a powder-based co-polyester coating, another type of plastic. Are the new bottles really safer than the old ones? Only time will tell.[67] My advice is to skip aluminum bottles altogether. And if you already have one, repurpose it for something else. One blogger wrote that after getting over her anger at being mislead by SIGG, she drilled a couple of holes in the bottom of her old SIGG bottle and used it as a flower pot.

Reusable Plastic Sports Bottles As I mentioned, I don't eat or drink out of any plastics these days (except for occasional sips from my travel mug lid). The plastics industry is simply too secretive about the chemicals they add to their products. Many reusable water bottles in the past were made from hard polycarbonate plastic which contained BPA. Now, many bottles are advertised as BPA-free, but as I mentioned in chapter 1, the jury is out as to whether the new chemicals replacing BPA are any safer.

Don't Reuse Disposable Bottles People often tell me that instead of buying bottled water, they simply refill their old disposable bottles over and over again. While this practice might be gentler to the environment, it is not good for our health. Plastics break down with exposure to light, heat, and rough treatment, causing any chemicals in them to leach out more readily. What's more, disposable bottles are difficult to clean,

Cleaning Your Water Bottle

Just as with plastic bags, it's important to keep reusable water bottles and mugs clean. Several water bottle companies sell special long brushes especially designed for their bottles; however, these brushes are plastic. If you already have a bottle brush, use that. But if you don't have one, choose a brush made from natural materials. My bottle brush is made from coconut fibers attached to a metal wire body with a wooden handle. I found it at a natural food store in my area, but you may also find them online. Currently, the web shop GetnGreen.com carries them. Or you can just search Google for "coconut coir bottle brush." I'll talk more about household cleaning supplies in chapter 7.

Clean reusable bottles and travel mugs with a natural fiber brush.

so bacteria can build up in them. Go ahead and toss the bottles you already have in your recycle bin. Then, decide on a durable bottle or container, and start using the heck out of it.

Choosing a Water Filter

Now you've got your bottle. How about the water you put in it? For years, Michael and I used a Brita water filter system to remove chlorine and any other contaminants that might be in our tap water. We hadn't really done any research but simply chose this brand because it was popular and inexpensive. A few weeks into my plastic-free project, it was time once again to replace the old filter cartridge. I remember standing in my kitchen holding the used up filter in my hand and realizing that here was one more hunk of plastic I would have to add to my collection. What was I supposed to do with it? Was there some way to recycle it? I went online and searched "recycle Brita filter" and found a website explaining that Brita had built a facility where spent cartridges were taken back and all the components recycled. Customers just needed to return the filter to the Brita bin at the store where it had been purchased. I got excited. And then I realized that the facility was in Germany and that the program only applied to Europe.

A little more research revealed that the North American branch of the Brita company had been purchased by The Clorox Company in 2000, and Clorox had not provided any way to recycle Brita filters in the United States. Why not? If it could be done in Europe, it could be done here too, right? In the months that followed, I started a campaign to convince the company to take back and recycle Brita water filter cartridges. That campaign was a success—Brita now provides a take-back recycling program for spent pitcher filters. I'll describe the campaign in more detail in the next chapter and explain why taking the cartridge apart and tossing it in my recycle bin was not an option. But early on, I just wanted to find a way to filter my water without generating more plastic waste. So I set out to find a water filter company that offered a recyclable or refillable cartridge. And I came up with nothing. Every company I talked to said to just toss the old filters in the trash. After all, that bit of plastic is tiny compared to all the plastic bottles you're saving by drinking tap water instead of bottled.

Still, I was committed to reducing my plastic waste as much as possible, so I started to wonder . . . did I even need to use a water filter in the first place? This was the question I should have asked from the beginning! Water systems vary throughout the United States and the world. Different contaminants can get into our water depending on where we live. People in agricultural states might have pesticides in the water. People in cities with old pipes might have more lead. Some municipalities use chlorine to purify the water while others use chloramine. People with well water have a whole different set of problems. And there are several different kinds of water filters which remove different kinds of contaminants. Finding out what's in the water is the first step in determining what kind of filter to get and whether you need one at all.

So Michael and I got our water tested and, after looking at the results, made the decision to drink our water straight from the tap without filtering it. No more plastic water filter cartridges for us. But this choice is not for everyone. As I said, water quality varies geographically, and not everyone is willing to ingest the chlorine, fluoride, and other chemicals added to the water by the provider. Here then, are the steps to take to figure out what kind of water filtration or purification you need:

1. If your water comes from a municipal system, first check your water provider's Consumer Confidence Report to find out what contaminants have been detected in your water. Visit the EPA's Local Drinking Water Information page to search for reports for your area. Or go directly to the web page for your water district. If the report is not posted online, you might have to contact the provider and ask for it.

2. After consulting the report for your area, you should also test the water that comes from your tap because contaminants like lead can leach directly from the pipes in your house or even from your city's system. Order a test kit from a certified lab. I used National Testing Laboratories to test my water.

3. Use the information in your Consumer Confidence Report and your personal water test results to decide what contaminants, if any, you want to filter out of your water. Different types of systems filter out different contaminants. The Environmental Working Group provides a fantastic online Water Filter Buying Guide, which explains the different types of filters and allows you to search the database of filters based on filter type (pitcher, faucet, under sink, on counter, whole house, and other types.), filter technology (e.g., carbon, reverse osmosis, ceramic), and specific contaminants.

4. Remember that a water filter cartridge is still one more piece of plastic. Yes, it takes the place of many, many plastic bottles. But what happens to it when it's used up? Can it be refilled or recycled? Generally no. Right now, the only water filter companies I know of with recycling programs are the pitcher systems from Brita, Zero Water, and MAVEA. If one of these systems is not right for you, then contact your chosen water filter company and ask them to create a way to recycle the filters. If the company tells you that you can just take it apart and put it in your recycle bin, be skeptical. Not all communities recycle all plastics, especially

oddly shaped plastics. You'll learn more about what can and can't be recycled in the next chapter. The point is that we don't get what we don't ask for. So speak up.

5. If, after testing your water, you realize that the only problem(s) you need to address are chlorine, taste, or odor, there are other options that involve less plastic. The Soma filter system uses a glass pitcher instead of plastic to hold the water, and the filter is encased in a plant-based plastic. According to the company, the filter is compostable in a commercial composting facility; however, keep in mind that not all commercial composters can actually compost all bioplastics. You'll learn more about bio-based plastics in Chapter 5. As of this writing, the Soma filter is only certified for reduction of chlorine, taste, and odor. If those are your only requirements and you don't care about certification, an even greener option might be Kishu. Kishu is advertised as the only plastic-free water filter. It's simply a stick of hand cut Japanese activated charcoal that comes packaged in a clear sleeve made from wood pulp. I hesitate to recommend it for more than taste and odor improvement because it doesn't have official certifications from an accredited testing lab, but if your water is otherwise fine, why not try it out? The carbon is compostable and there's no plastic to worry about.

Is Your Water Cooler Messing with Your Hormones?

Installing a water cooler might seem like a good way to have your bottled water without generating a ton of plastic bottle waste. Many offices, for example, have water delivered in 5-gallon jugs that are swapped out with each new delivery. The jugs are refilled and reused multiple times, with the only waste being the new plastic cap on each container. But while these systems do reduce a lot of plastic waste when compared with buying cases of bottled water, the truth is that they are still another form of bottled water. The jugs are problematic, since they are often made from polycarbonate plastic (#7 PC) which contain BPA (although some companies have switched to PET recently). What's more, the water itself is no better regulated than any other bottled water product, and it requires fuel to ship. A better

> "I use a big quart jar to keep tap water at work so I avoid frequent trips to the break room water cooler." —Kent Lewandowski, Healthcare IT consultant and Sierra Club activist, Oakland, California

option is to install a bottleless water cooler that connects directly to the plumbing and eliminates the need for any water to be delivered in the first place. Examples are Quench, iBottleLESS Water Solutions, Culligan Bottle-Free coolers, Blue Reserve, Waterlogic, and others that purify, heat, and cool your own tap water and don't require lifting a bunch of heavy jugs. At the big corporate law firm where Michael works, the office administrator and the administrative partner decided that it was a waste to keep the refrigerators stocked with bottled water, so they had several filtered water systems installed on each floor, and issued every employee, from partners to file clerks, a 40-oz. Klean Kanteen. For client meetings, they put out pitchers of ice water. It can be done!

Another alternative, if you don't want to install a water filter or bottleless water cooler in your home, is to bring your own jugs to fill up at grocery stores that sell filtered local water right from the source. Companies like U.S. Pure Water and FreshPure Waters provide bulk water filtration machines to natural foods stores, using technologies like reverse osmosis and/or deionization to purify the local tap water. Customers can fill their own jugs for a fraction of the cost of bottled water. For example, my local Whole Foods Market in Berkeley sells bulk filtered water for 49 cents per gallon. These stores usually offer plastic jugs for sale, but you can bring your own glass containers to avoid the plastic. Beer and wine brewing supply stores offer glass carboys in sizes ranging from one to six gallons. Google "glass carboy" to check prices, or look for used containers via Freecycle or Craigslist.

Beyond Bottled Water: Fizz & Flavor

Once I had kicked the bottled water habit, I started noticing all the other ways that water is bottled in plastic and shipped around the world. What is soda but bottled water with fizz and flavor? Energy drinks, sports drinks, special elixers like Vitamin Water or Sobe Lifewater are just bottled water with extra stuff added. What if we could make our own? We can!

Add Carbonation. If you drink carbonated beverages on a regular basis, it might be worthwhile to invest in a soda maker and carbonate your own tap water. But keep in mind that some soda makers are more wasteful than others. Soda siphons add carbonation to the water via a small single-use stainless steel "charger," which may or may not be recyclable where you live. Each charger only holds enough carbon dioxide to make 1 liter of soda. This system simply replaces single-use plastic bottle waste with single-use stainless steel waste, which has an even higher carbon footprint! A less wasteful soda maker is one which adds CO_2 via a returnable, refillable canister. Once the canisters are spent, they can be exchanged for new ones at participating retailers, which take back the old ones to be refilled. Both Sodastream and Cuisinart offer these kinds of machines. While the Cuisinart Sparkling Beverage Maker includes a reusable plastic bottle in which to make the soda, Sodastream offers several models that come with glass carafes instead of plastic. However, as of this writing, there is an international boycott of Sodastream protesting the location of its factory in occupied territory in the West Bank. This kind of political issue is beyond the scope of this book. You can learn more about the Sodastream boycott and alternatives to Sodastream at GlobalExchange.org. Type "Sodastream" into the search box. Maybe, after weighing the pros and cons, you'll decide that owning a soda machine is not necessary in the first place. After all, soda machines are made from plastic, and like any appliance, they require energy to manufacture. A soda maker is a green option if used regularly to avoid plastic bottled sodas. It's not so green if it sits on the shelf collecting dust.

Soda stream Penguin home soda maker

Brew It Yourself. In the old days, sodas were made via the fermentation of yeast, which is a much more involved process than pumping CO_2 from a canister but can be a fun home brewing project. Google "How to make homemade soda" for instructions. Mother Earth News has a fantastic online article on making homemade root beer, ginger beer, and grape soda the old-fashioned way To find the article, Google "Mother Earth News Homemade Soda."

> "I take my reusable bottle or mug just about everywhere I go. If I want 'herbal' water, I make it myself. But I call it tea." —Diane MacEachern, author, www.biggreenpurse.com

Add Some Flavor. When I was a kid, my mom refused to buy orange soda for us because we had powdered Tang at home and could make our own orange drink. If it was good enough for the astronauts, it was good enough for us! Mom would have been proud of Michael, who gave up buying bottled Gatorade a few years ago when he discovered he could save money by buying Gatorade powder (on the same shelf next to the Tang, Country·Time Lemonade, and Crystal Light) and mix his own sports drinks. While the powder comes in a plastic container, that one container can make several gallons of drink, depending on how strong you mix it. Me? I'm not such a fan of sweet chemically concocted drinks these days, opting for a natural squeeze of lemon, lime, or a few slices of cucumber in my water. But for those who are more adventurous, flavoring and supplement options abound. Companies like Torani and Williams-Sonoma bottle gourmet soda syrups in glass. And the Internet is full of recipes for making your own sodas, teas, and infusions, as is Andrew Schloss's book *Homemade Soda: 200 Recipes for Making and Using* (2011, Storey Publishing), which contains a wealth of ideas for enhancing water, including recipes for fruit sodas, fizzy juices, root beer and colas, herbal waters, sparkling teas and coffees, cream sodas and floats, and even energy drinks. All of the recipes in the book could be adapted for either still or carbonated water.

Save Glass-bottled Sodas for Special Occasions. I love Reed's Extra Ginger Brew ginger beer. It comes in a glass bottle. It's super spicy and delicious. And every once in a while I like to treat myself. I don't buy soda every day because even though glass is nontoxic and recyclable, it requires tremendous amounts of energy to produce and ship, and I can make ginger soda with my own tap water, soda maker, gingerroot, and sugar. Still, being green doesn't have to mean being deprived. The website GlassBottleSoda.org is a great resource for finding sodas bottled in the United States and packaged in glass. Some of the companies listed still take back their glass bottles and refill them.

Hunting for Drinking Water

Standing in line with my friend Laura at the entrance to the Outside Lands Music Festival in Golden Gate Park in August, 2008, I didn't want to think about plastic or the environment or sea turtles or albatrosses. I was there to experience the exquisite noise of my favorite band, Radiohead, and let myself melt into the music. But it would be hours before the band went on, and in the meantime, I'd stroll through the park, eat some gourmet food, drink a little wine, listen to other bands, and absorb some late summer sun. I'd come prepared with a reusable stainless steel water bottle, filled from my tap before I left the house, and since this festival had been billed as a "green" event, I was ready to just relax and have fun. I was not prepared to be confronted by the guard at the gate demanding that I empty my bottle before entering.

"You can't bring liquids in here."

"But it's just water."

"I don't care what it is. It's our policy. You have to pour it out."

"Will I be able to refill it inside?"

"Yes."

It seemed like a waste of perfectly good water, but we didn't want a hassle, so Laura and I emptied our bottles onto the ground, entered the gate, staked out a spot near the stage, and then I set off to refill our bottles.

Spotting a couple of guys manning one of the many recycling stations in the venue, I held up my bottle and asked, "Where's the water?"

"We don't know," said one guy. "Let us know when you find out."

I kept walking and spotted a first aid station. "Where's the drinking water?" I asked.

"Hmm . . . good question. Not here."

I walked what felt like the length of twelve football fields, passing vendor after vendor selling beverages in plastic bottles until I spotted a big sign up ahead: WATER. But as I got closer, I could see that this "water station" too only offered single-use plastic Arrowhead and Aquafina bottles, and for an exorbitant price. "Where's the water to fill my bottle?" I asked. The water sellers just shrugged.

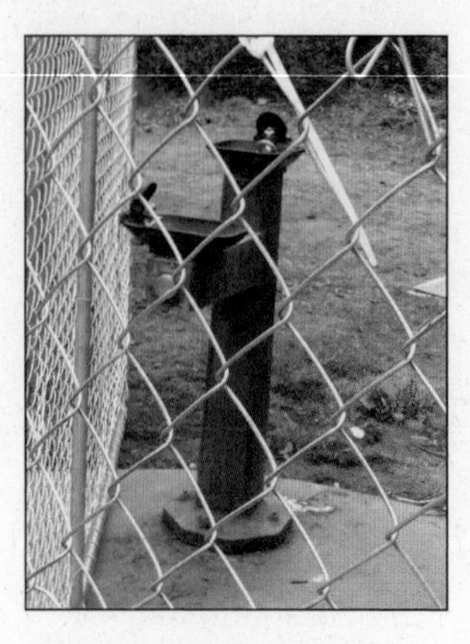

Public fountain off limits at Outside Lands Music Festival

It was getting late. Laura would be wondering where I was. I was so determined to fill up our bottles, I was ready to use the water from the hand washing station near the portable toilets. Fortunately I saw the sign first: THIS WATER IS NOT FOR DRINKING.

And then I noticed something real: an actual drinking fountain right near an actual restroom. But where was it? Behind a tall chain-link fence, blocked to festival attendees! An official-looking guy in a blue uniform walked by. I asked him if I could get through to fill up my bottle. "Of course not," he snorted. "We have no intention of providing free water to everyone at this festival. I don't know why you'd think that." Um . . . maybe because that's what they told me at the gate when they made me pour out the perfectly good tap water I'd brought with me? And because I paid big bucks for a ticket? And because this is supposedly an eco-friendly event?

Laura and I did end up getting our bottles filled that day. A few minutes after blue uniform man walked away, a guy in board shorts trudged up to the fence carrying a big jug. He picked up a hose lying on the ground, a hose that ran from inside the restroom along the ground and through the fence, and started filling his container. It turns out that hose was for the coffee vendors and had been lying right there all along. I told coffee guy my saga, and he filled our bottles with fine San Francisco tap water.

Radiohead rocked that night. Thom Yorke took my breath away. "A fake plastic watering can," he crooned, "for a fake Chinese rubber plant. In the fake plastic earth . . ." And then ninety minutes later, the show was over. Stark white light flooded the field, revealing the ugly reality of thousands and thousands of crushed plastic bottles strewn across the grass as tired attendees kicked them aside, heading for the gates. I went home that night and wrote a letter about bottled water to the festival organizers. And a blog post. I think I wrote a letter to Radiohead too, you know, while I was in letter-writing mode and all.

Apparently enough other attendees spoke up as well, because in all subsequent years, Outside Lands Music Festival has provided water bottle filling stations for people who bring their own bottles and sold reusable bottles for people who don't. They still don't provide tap water, choosing instead to truck in those 5-gallon polycarbonate bottles, but it's a step in the right direction.

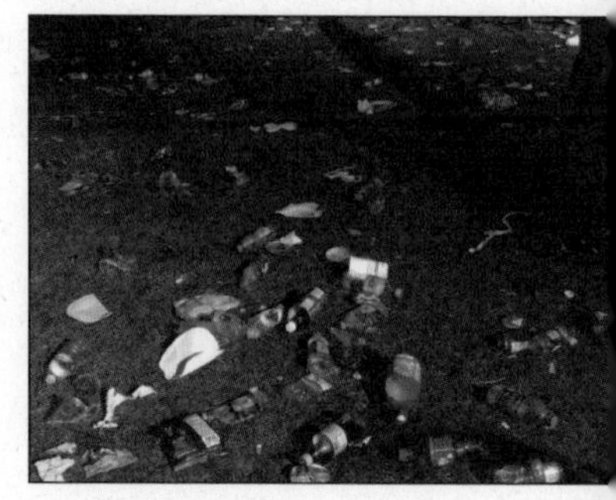

Plastic bottles litter Outside Lands Music Festival at the end of the night.

This story highlights an ongoing obstacle to using our own bottles. It's fine for us to fill up at home, but what happens when we're out in the world and need a refill? How do we find clean plastic-free drinking water? Quite a few people are thinking about this issue and working to create practical solutions.

Online Tools

- **TapIt Water** is a website to help us find cafés and restaurants that will fill our reusable bottles while we're out and about. Restaurants can sign up on the site as partners, and users can check the map or download the free mobile app. While the program began in New York City, the network is expanding across the United States. But even if you can't find a café listed on the TapIt site, it never hurts to just walk into an establishment and ask. Even better, fast food restaurants and convenience stores like 7-Eleven with DIY soda filling stations generally provide one dispenser for free tap water. My strategy? Dispense first; thank later; ask not at all. It's just easier that way.
- **WeTap** is a project to map and improve our drinking fountain infrastructure and educate the public on the safety of tap water. The project includes a Google map of water fountains in major U.S. cities, which can be accessed via any computer or smartphone with web access, as well as a mobile app, Users can upload information about the location and condition of public drinking

PUBLIC WATER HERO:
Evelyn Wendel, founder of WeTap

After reading Charles Fishman's eye-opening article, "Message in a Bottle,"[68] about the marketing of bottled water, Evelyn Wendel committed to doing something to solve the bottled water epidemic. She bought the domain name WeTap.org and started brainstorming solutions, including ideas for how to make water fountains convenient and accessible again. There were so many ways to go: build new ones, fix old ones, or redesign them altogether. What she realized was that before anyone could upgrade the water fountain system, they needed to know where all the fountains were in the first place, and in her city of L.A., there were no such records. She threw her energies into mapping water fountains in the L.A. area, beginning with the UCLA campus, and with the help of some UCLA students, developed the first WeTap mobile app. In 2010, Evelyn teamed up with water expert Dr. Peter Gleick, who has since become a consultant and adviser to WeTap. Together, they have developed updated versions of the drinking fountain app for Android and iPhone and have expanded the project to include other cities in the United States and around the world.

Evelyn is someone who knows how to speak up and never misses an opportunity to educate others on the environmental problems created by bottled water. While she was watching an episode of the Disney TV show *Good Luck Charlie* with her eleven-year-old daughter, Maude, Evelyn noticed a scene in a grocery store with a lot of plastic water bottles. Coincidentally, Evelyn knew Drew Vaupin, the producer of the show—their kids went to the same school together! She had to say something. And as it turned out, Drew needed very little convincing. After several conversations, the producer agreed to not only ban bottled water from the set of the show but to also eliminate it from the show itself. In a message to WeTap, he wrote:

"Our show has implemented a no-plastic bottle policy for all future episodes that are produced. If you see a character use one or spot a bottle in a soda machine, it's from Season One and we apologize. Thanks for bringing our attention to this matter."

Speaking up can make a big difference. "What we choose to think about makes a difference," Evelyn says. "Not everyone goes to school with Disney producers, but everyone goes to markets, restaurants and makes personal choices." There are many different opportunities for each of us to speak up and make a difference. "All of our voices matter. All of our actions count."

fountains in their communities. You can even upload photos of drinking fountains.

- **Find-a-Fountain** is an interactive site and mobile app mapping the location and condition of drinking fountains in London.

Hydration Stations If you've ever tried to maneuver your water bottle under a typical water fountain spigot, you know how awkward it can be. There's not enough room under the faucet to keep the bottle upright, so you end up with a half-full bottle and a lot of wasted water. But innovative companies are cropping up to address that need with machines designed to allow quick and efficient bottle filling with less mess. Machines range from simple devices that dispense plain water to those that add flavors and even fizz. Here are a few examples. Consider whether any of these machines might be appropriate for your city, organization, or event.

- **GLOBALTAP** water refill stations shoot water straight down into your bottle. The City of San Francisco has begun installing these devices in outdoor public spaces throughout the city and at San Francisco International Airport to encourage visitors to drink free tap water instead of bottled. Visit drink.sfwater.org to find out where the SF bottle filling stations are located.

Quick Tip:

Bring your empty water bottle with you through airport security and fill it up from the water fountain inside the secure area of the airport. You *can* bring your own water on a plane; you just can't bring it through the security gates.

- **Elkay EZH2O** touchless bottle filling stations work as standalone devices or in combination with traditional drinking fountains. They dispense chilled water straight into your bottle, while a "green ticker" keeps track of the number of disposable bottles saved from the landfill. Machines are available with or without filters. In 2009, the students of the Brentwood School (K–12) in Los Angeles, California, initiated a water bottle project to stop the waste from water bottles on their campus. As part of the project, the school installed five Elkay EZH$_2$O machines and stopped providing bottled water. To encourage students to use the machines, they designed a Brentwood School reusable water bottle that they sold at a discount through the bookstore, handing them out for free to the elementary school kids. According to Will Bladt, the middle division director, the program has been a huge success. The kids much prefer filling their own bottles from the new fountains, and in less than a year, the program saved over 65,000 bottles from the landfill.[69] In September 2011, *USA Today* reported that similar programs are popping up at hundreds of college campuses across the country.[70]

It's okay to take your empty bottle through airport security and fill up at the water fountain inside the gates.

- **Brita Hydration Station** combines Brita filter technology with Haws drinking fountains to create touchless bottle filling stations. To promote filtered tap water over plastic bottles, Brita's Filter For Good campaign has been teaming up with

Hydration Station at San Francisco International Airport

popular musical acts like Black Eyed Peas, Dave Matthews Band, Jason Mraz, and Jack Johnson to provide water refill stations at concert venues where bottled water is typically sold.

Bottled Water Legislation

In 2008, the U.S. Conference of Mayors adopted a resolution encouraging cities to phase out, where feasible, government use of bottled water and promote the importance of municipal water.[71] San Francisco's Mayor Gavin Newsom had already cancelled bottle water contracts at City Hall, and mayors from Boston, Chicago, Philadelphia, New York, Miami, and many other cities have followed suit. Now, campaigns are underway to encourage state governors to cancel bottled water from their offices as well. Also in 2008, Toronto banned bottled water from all municipal property. But unlike the plastic bag measures passing with increasing frequency in cities throughout the world, most bottled water actions stop short of outright bans from private businesses.

In September 2009, the tiny town of Bundanoon in New South Wales, Australia, became the exception. In an initiative called "Bundy on Tap," the city's businesses and residents voluntarily agreed to ban prepackaged single-use bottled still water from all shops in the town. Instead of offering bottled water, stores now sell reusable bottles; the town has installed public water stations; and some local businesses provide free chilled, filtered tap water. What prompted such a bold action? For several years, residents of Bundanoon had been actively opposing a nonresident corporation that sought to drill a borehole and extract water from the community for commercial purposes. One local businessman, Huw Kingston, came up with the idea that if the town were truly opposed to outside interests extracting its water, it ought to also oppose the end product of that extraction—bottled water itself. He published a letter to the community in a local paper, and quickly garnered support from both residents and merchants, who voted to become the world's first Bottled Water Free Town.[72]

The following year, a similar measure passed in the tiny town of Concord, Massachusetts, The story on page 108 of Concord's bottled water ban is an inspiring lesson for any of us who want to make a difference in our communities.

Take Action in Your Community

To learn more about issues around bottled water or take action in your organization or community, check out these helpful resources.

Organizations

- **Ban the Bottle** is a comprehensive website dedicated to banning bottled water and a great resource for anyone wanting to start a campaign.
- **Corporate Accountability International's "Think Outside the Bottle" Campaign** asks supporters to sign a pledge to refuse bottled water and works against the privatization of this essential resource.
- **Food & Water Watch's "Take Back the Tap" Campaign** promotes tap water and works to protect our drinking water supplies.
- **Polaris Institute's "Inside the Bottle" Campaign** provides information for schools and municipalities that seek to ban bottled water.
- **Bundy on Tap** is the official website of the campaign to ban bottled water in Bundanoon, NSW, Australia. Read how they did it and get inspired.

Useful Articles

- **"Message in a Bottle: Despite the Hype, Bottled Water Is Neither CLEANER nor GREENER Than Tap Water,"** by Brian Clark Howard in *E—The Environmental Magazine*, Aug. 31, 2003.
- **"Message in a Bottle,"** by Charles Fishman, in Fast Company, July 1, 2007.

Books

- ***Bottlemania: How Water Went on Sale and Why We Bought It***, by Elizabeth Royte (Bloomsbury:2008).
- ***Bottled & Sold: The Story Behind Our Obsession with Bottled Water***, by Peter H. Gleick (Island Press:2010).

Films

- **Tapped** Stephanie Soechtig's relentless exposé examines the inside story of the bottled water industry and travels from the factories where the plastic bottles are produced to the ocean where much of the plastic finally ends up and all the places in between.
- **Blue Gold: World Water Wars** Narrated by Malcolm McDowell and based on the book *Blue Gold: The Fight to Stop the Corporate Theft of the World's Water*, this film follows numerous worldwide examples of people fighting for their basic right to water.
- **FLOW: For Love of Water** A comprehensive look at water issues around the world, showing the struggles that communities from Michigan to India are undertaking to protect their most precious resource—water.
- **Divide in Concord.** The irreverent story of Jean Hill's campaign to ban bottled water in Concord, MA.
- **The Story of Bottled Water** Annie's Leonard's follow-up to her popular video, The Story of Stuff, looks at the bottled water cycle and how companies convince us to buy their product. View it at StoryofStuff.org or on YouTube.

PLASTIC-FREE HERO:
Jean Hill

Haunted by the images of the plastic pollution swirling around the North Pacific Gyre, a phenomenon she learned about from her grandson Mac, octogenarian Jean Hill decided she had to act. With more energy and commitment than many people a fraction of her age, Jean started to research the politics of bottled water and the political process of her small town of Concord, Massachusetts. She wrote a petition article (similar to a ballot measure in other parts of the country) banning the sale of bottled water in Concord and gathered the necessary signatures to get it posted in the Town Warrant. With the assistance of her friend, Jill Appel, she held a screening of the anti-bottled water documentary ***Tapped***, which she announced in the Concord Journal, so readers would know the time and date. Afterwards, she and many friends and supporters of the bottled water ban wrote letters to the ***Concord Journal***, and Jean contacted organizations like Food & Water Watch and Corporate Accountability International for information she could use to put together an informational flyer. Jill and her son created a PowerPoint presentation to show at the annual Town Meeting, where the city's residents come together once a year to vote on legislation by show of hands. In May 2010, the measure passed. But shortly thereafter, the state's Attorney General disallowed the bottled water ban because it did not include any provisions for enforcement.

Undeterred by this first setback, Jean got to work revising her bottled water ban to present at the town meeting the following year. The new measure would be enforced by quarterly inspections by the public health department. There would be graduated fines for the first and subsequent infractions, fines which would be set by Concord officials and paid to the Concord town treasury. In April 2011, Concord residents met once again to do their civic duty. But in the year since Jean's first ballot measure passed, representatives of the bottled water industry had actively campaigned against it. Residents feared a lawsuit if the measure passed. The bottled water ban was defeated by a margin of seven votes. Instead of the ban, members approved a resolution to educate citizens on the environmental dangers of bottled water. Jeff Wieand, chairman of the Board of Selectmen, told the ***Boston Globe***, "The cost of defending ourselves against such a lawsuit could be steep. It's possible we could get a law firm to defend us pro bono, but if that didn't happen it would be a significant expense for the town."[73]

Jean thinks there was another reason the measure failed. She says the article wasn't introduced until late in the evening after many of the meeting's attendees had gone home. She is not giving up. "I'm going to get the damned thing passed," she told me emphatically during a phone conversation a few months later. "If my next attempt makes it through, I'll be eighty-five by the time it goes into effect, but I'm not stopping." And then she added, "I believe every citizen has a moral obligation to do something for their town, their country, and their planet." Jean lives her beliefs every day. Her persistence in the face of obstacles inspires me to keep learning and keep trying.[74]

On April 25, 2012, after the first edition of *Plastic-Free* went to press, residents of Concord, Massachusetts passed Jean's measure to ban bottled water in the City of Concord. The ban took effect January 1, 2013, and to date, challenges to the ban have been resoundingly defeated.[75]

Action Items Checklist

(Choose the steps that feel right to you. Then, as an experiment, challenge yourself to do one thing that feels a little more difficult. Only you know what that one thing is.)

- ☐ Obtain a reusable beverage container.
- ☐ Make a plan to carry/remember your reusable bottle/mug.
- ☐ Practice bringing your reusable bottle/mug with you when you leave the house.
- ☐ Check the Consumer Confidence Report from your water provider and/or get your tap water tested to see if you need a water filter.
- ☐ Purchase a water filter, if necessary, and drink only tap water at home.
- ☐ Talk to your friends and family about why you have decided to stop buying bottled water. Explain, but don't nag! Seriously.
- ☐ Work on getting bottled water out of your workplace, school, church, or other organization. Provide information on bottleless water coolers or hydration stations.
- ☐ Watch the film *Tapped* alone or with friends.
- ☐ Organize a screening of the film *Tapped* through a civic group, church, PTA, or other organization.

- ☐ Talk to local store managers about setting up bulk water stations or bottle refill stations and eliminating bottled water from store shelves.
- ☐ Write a letter to the editor of a local paper explaining the problems with bottled water.
- ☐ Write to a celebrity who endorses bottled water and ask them to stop. Explain the reasons why.
- ☐ Write to your legislators about cancelling bottled water contracts or banning bottled water.
- ☐ Join up with Corporate Accountability International or Food & Water Watch to learn what else you can do and take action.

Chapter 4: Why Can't We Just Recycle It All?

Before I tell you more about my adventures in de-plasticking, I want to talk a little bit about recycling. Why? Because throughout this project, when I have explained to people about how I don't buy new plastic, I have often heard the response, "But this bottle or container or bag or product is recyclable, so that makes it okay to use." And while it's true that more and more plastic recycling companies are popping up to provide second lives for many of the products and packaging we discard, recycling has drawbacks. First, plastic recycling doesn't address the toxicity issues associated with plastic. But beyond that, plastic packaging is usually downcycled into secondary products that are rarely if ever recycled themselves; there are economic limitations to plastic recycling; there are environmental issues associated with the recycling process itself; and the availability of recycling systems can lead people to believe that it's okay to continue buying disposable plastic.

Back in 2007, before I understood anything about plastic recycling, I was frustrated that my city of Oakland only accepted "narrow-necked bottles," and I couldn't understand why I had to lug our cottage cheese and pudding containers to my friends' recycle bin in San Francisco. I was the bag lady on the BART commuter train toting trash from one side of the Bay to the other. And I, like most people, was under the impression that "recycling" simply meant tossing my trash into the recycle bin and forgetting about it. It wasn't until I visited some actual recycling facilities and spoke with recycling professionals that I learned the truth.

Curbside Recycling: A Visit to the Davis Street Municipal Recycling Facility

Anxious to learn what really happened to the materials I placed in the recycle bin, I made an appointment for a tour of my local materials recovery facility, or MRF (pronounced "murph"). Rebecca Jewell, program manager for Waste

Management's Davis Street Transfer Station was happy to show me around. After outfitting me with a hard hat and reflective vest, she led me out to an immense building, Oakland's new $9 million facility, where 400 tons of the Bay Area's recyclable waste was processed every day. Waste Management trucks arrived one after the other, dumped mixed up piles of paper, cardboard, glass, metal, and plastic (we have a single-stream recycling system) on the tipping floor, and then drove off to collect more. The piles were then loaded onto the pre-sort line conveyor belts that sped by at 50 feet per minute, as employees removed obviously nonrecyclable tems—including wood, e-waste, and anything in plastic bags—from the process.

Davis Street Materials Recovery Facility

Past the initial sort, various technologies separated the materials mechanically. Large fans blew lighter materials like paper one way, while heavier objects like containers fell and were directed another way. A metal screen broke all of the glass. The heavier materials then passed under powerful magnets, which pulled out tin and steel cans. Employees stood by at the other end of the line, each one assigned to look for and manually remove a specific type of item—PET bottles, HDPE jugs, etc. Workers had to move quickly and identify items as they passed by based on color and shape alone. There was no time to check the bottoms for the plastic identification number. Lastly, a reverse magnet caused the aluminum cans to fly off the conveyor belt into a bin. At the other section of the sorter where the paper ended up, another group of employees manually separated out the various kinds of paper and cardboard as they went whizzing by. Each type of material was compressed and bundled into bales for shipping. If a bale contained more than 10 percent contamination, the entire batch was ruined and couldn't be sold.

Um . . . actually, none of the stuff I just described was happening when Rebecca and I entered the MRF. The building was surprisingly quiet. Climbing up the metal steps to see what was going on, we discovered workers struggling to remove strappy film out from the teeth of a sorting machine. Anything long or filmy will cause it to cease functioning. Think of your vacuum cleaner. You know how it jams up if you suck up a piece of string or cord and how you then have to spend time unwinding it? A sorting machine gets jammed in the same way, only on a much bigger scale. It's a headache for workers if they don't catch those materials and remove them before they make it to the mechanical sorter. That's why Oakland doesn't allow plastic bags or any kind of sheeting in its curbside recycling program.

Leaving the MRF that day, I understood why so few plastic items were accepted in our program. With so much material going by so quickly and with humans racing to pull out the most obvious recyclables, it would be impossible to identify and recover everything that might be theoretically recyclable. In the few years since my little field trip, technology has advanced a lot. Davis Street now employs optical sorting devices that scan plastics using infrared radiation and can separate them by type and even color much faster than human eyes or hands. But not all recycling facilities are so high-tech, which is why many municipalities only accept jugs, bottles, and tubs that are easy to spot.

We exited the MRF through the back, where Rebecca showed me what happens to the baled materials after they are processed. Just outside, a long line of empty container trucks waited to be filled up. Most of them would head to the Port of Oakland, where the containers would be loaded onto ships bound for China and other Asian countries. The truth is that most of the plastic recycling from the United States and much of the world is shipped overseas. There are some plastics reclaimers (companies that grind, wash, and convert the material into pellets or flakes that can be used again in new products) in the United States, mostly in the Southeast, but the greatest market is in China, where so many plastic products are manufactured in the first place. There's a certain economic sense here. After receiving new plastic products from China, we fill up those container ships with our plastic recycling rather than sending them back empty.[76] Yes, our biggest export to China is trash.

I'll talk more about what happens to plastic in China, but first, here are the helpful rules I learned from my trip to the recycling center.

General Rules for Plastic Recycling

A Triangular "Chasing Arrows" Symbol and Even the Word "Recyclable" on an Item Does Not Mean That It Can Be Recycled! I already mentioned this point in chapter 1, but it's worth repeating. Municipalities and recycling centers make rules about what types of plastic they will accept for recycling based on market demand for the material and their ability to sort it. For example, even though #1 PET is one of the more recyclable types of plastic, few facilities will accept those black plastic #1 food trays that frozen meals come in. Yet they are advertised as recyclable because theoretically, they could be. But if there is no recycling market for the item, and if your city will not take it, then it's not really recyclable, is it?

Be Sure You Know Not Only What Plastic Numbers Are Accepted by Your City, but Also What Shapes of Items. Oakland accepts narrow-necked bottles, regardless of the number, and #2, #4, and #5 tubs and lids, the types of containers that hold yogurt or cottage cheese or various spreads. But the city will not accept, for example, my little eye drop vials with a #4 on them. Different shapes of containers have different properties, even if they are made from the same kind of plastic.

Beware of Bottle Caps. Some facilities will not recycle the bottle with the cap left on. Some will. Either way, caps left on bottles will trap air, making the bottles harder to compress, and may also trap liquids. Find out what your city's policy is. Remove caps if necessary and send them to an alternative bottle cap recycler, which I'll describe further in this chapter.

Food Containers Should Be Rinsed Out. You don't have to scrub with soap. But especially in single-stream systems, food left in containers can contaminate paper and render theoretically recyclable materials useless. What's more, it's common courtesy. Patty Moore from Moore Recycling Associates in Sonoma, California, wrote to me: "Please let people know that once they put that plastic container in the bin, human beings are going to deal with it (both here in the United States and in China if it gets exported), so please take the extra time to scrape off and rinse off first."

Cardboard Milk Cartons May or May Not Be Accepted for Recycling. Cardboard cartons for milk, juice, and other foods are coated inside and out with plastic (not wax!). Some cities accept them, rinsed out, with the paper recycling. Other cities may accept them in a compost bin. Either way, they require special handling. Find out your city's policy. And read the next chapter to learn about how plastic-coated cardboard may be problematic for composters.

Be Careful Recycling Plastic Bags! Most cities do not accept plastic bags for curbside pickup because they can jam sorting machines. But in those that do accept bags, it's important to follow the procedures for proper disposal. *Do not* use them to hold your other recyclables and don't put them into the bin loose. If your city really does accept plastic bags in the bin, you should stuff several bags inside one bag and knot it closed so that they cannot escape. If your city doesn't accept plastic bags in the curbside bin, return them to the grocery store where you got them.

> "I routinely pull plastic bagged newspapers out of our condo recycling bins, bins that have a label that says no plastic bags. Since the bags that the newspapers come in can be recycled along with grocery bags, I hike them over to the local grocery store." —Clif Brown, Chicago, Illinois

Aseptic Packages Are Very Difficult to Recycle. Aseptic cartons are shelf-stable containers like juice boxes, wine boxes, soy and rice milk containers, or soup containers that allow normally perishable products to be stored unrefrigerated on grocery store shelves. Aseptic cartons are made of multiple layers: plastic, cardboard, aluminum, cardboard, plastic, and these materials are difficult to separate and process. Some cities accept them for recycling. Others do not. Those that do must send them to specialized recycling facilities, which are rare. Because of the difficulty of recycling them, it's best to avoid aseptic cartons when possible.

Don't Put Thermal Receipt Paper in the Recycle Bin. Most thermal receipt paper is coated with BPA, which will contaminate any products made from recycled paper. Recently, BPA has been detected in recycled toilet paper. Thermal receipts should go in the trash.

Just Because Your City Accepts Certain Items for Recycling Does Not Necessarily Mean That They Are Actually Recycled. One strategy that some municipalities use to increase recovery rates is to accept all plastics, #1 through #7. The idea is that if folks don't have to figure out which plastics to put in the bin, they will toss everything in, and the workers will separate out the items they can recycle. What doesn't get recycled gets landfilled or incinerated. Unfortunately, a side effect of this strategy is that consumers may be less conscious of their purchasing choices, assuming that everything they buy is getting recycled when it's not.

Never Place the Following Items in Your Curbside Recycle Bin. Animal waste, batteries of any kind (take them to special battery collection centers), footwear of any kind, electronic appliances, electronic toys, engine oils, fluorescent tubes and bulbs (they contain mercury), golf balls, green waste, household hazardous waste, liquids, soccer balls, stuffed animals, tennis balls, wigs, blankets, hoses, ropes or other strapping materials, bowling balls, bricks, concrete, engine parts, rock, tires, toilet seats, and other building materials. Sound ridiculous? These are examples of actual items that have found their way into curbside recycle bins in the Bay Area. The trouble is not just that they are not recyclable but they can actually damage the sorting equipment and injure the workers. Believe it or not, one of the biggest hazards in the recycle stream at Davis Street are mini propane tanks. The workers there had collected a whole bin of them. Care for an explosion, anyone?

The Business of Recycling

The only plastic that actually gets recycled is that which plastics reclaimers are willing to buy. Recycling is a business, and if there is no market for a material, it will end up in the landfill or incinerator. The most valuable types of plastic are #1 PET and #2 HDPE bottles, and yet of the 9.1 billion pounds of PET and HDPE bottle resin sold in the United States in 2013, only 31 percent was recycled.[77] If 31 percent weren't already a failing grade, the Container Recycling Institute estimates that the rate is actually even lower "because a quarter of the weight of collected PET bottles consists of caps, labels, glue, base cups and other contaminants—and is therefore unusable as reclaimed PET feedstock."[78] And although whole new industries are developing to recycle other kinds of plastics, the recycling rates for plastic films and non-bottle rigid plastics are even lower. There are various reasons for the low rates. Some plastics, like PVC or polystyrene, are less desirable. In fact, the Association of Postconsumer Plastic Recyclers (APR) considers PVC bottles to be a contaminant in the recycling stream.[79] Another reason is the cost of processing. While it's possible to incorporate recycled PET back into new bottles, cleaning the material to the standards necessary for food contact is much more resource intensive and expensive than simply using virgin resin, so much of it gets downcycled into secondary products like carpet or polar fleece. And the way recycling centers combine mixed plastics into bales impacts the quality of the material available for sale. The APR has created bale standards to help recycling centers process plastics in a way that is the most marketable, but the fact is that many communities don't have the resources to follow these guidelines.

Environmental Impact of Plastic Recycling

As I mentioned earlier, recovered plastic that is not sold domestically is shipped overseas, mainly to China, which has historically accepted lower quality plastic bales, although that is changing. In 2013, China instituted a policy called Operation Green Fence to require customs to strictly enforce laws governing the import of waste. Still, China does not impose the same environmental regulations or worker safeguards as the United States. In 2007, Britain's *Sky News* aired an exposé on the conditions in Lian Jiao, a Chinese town that had become a toxic waste dump for the West's plastic recycling. The video is shocking. Piles of plastic trash line the streets. Workers melt down plastic without wearing any kind of masks or protective gear. Toxic fumes pollute the air, and chemical wastes pour directly into the town's water supply, while children crawl through mounds of plastic trash.[80] Shortly after the program aired, the Chinese government closed down this particular facility.[81]

I've spoken with several plastic recycling advocates who assure me that situations like Lian Jiao are the exception not the rule, and that most plastic recycling operations in China, while not necessarily up to U.S. standards, are not as bad as that one example. But the fact remains that environmental standards in China are very low. A friend who returned from a visit to that country in fall of 2011 reported that she never saw the sun because of the thick smog blanketing the cities she visited and that she suffered a burning throat and itchy eyes for the entire week she was there.[82] What's more, air pollution produced in China doesn't stay in China. Every year, according to a report in *Discover* magazine, prevailing winds push tons of mercury, sulfates, ozone, black carbon, and desert dust from China over the Pacific Ocean, pollutants that can reach North America within days.[83] Just as manufacturing virgin plastic can create air and water pollution, so can the processes used to recycle the material if steps are not taken to reduce emissions.

RECYCLING HERO:
Lisa Sharp

For people like me, who live in the Bay Area, where curbside recycling is taken for granted, it's hard to imagine that there are still communities where locals have to load up their paper, glass, metal, and plastic in the car and haul it all to the local recycling center themselves. As you can imagine, only the most die-hard environmentalists are willing to make the effort. Lisa Sharp is one of those committed souls. She lives in the small town of Ada, Oklahoma, where some locals consider their trucks to be Magic Trash Cans: pile up all your loose junk in the back of the truck, drive down the highway fast, and by the time you get to the dump, it's all disappeared. Like magic. It takes work to educate the populace about recycling and waste.

So Lisa joined the board of the Ada Recycling Coalition, an organization that works together with the local Native American tribe, the Chickasaw Nation, as well as city managers to improve and increase recycling by providing education and referral services to Ada's residents. Setting up educational booths and providing trailers to collect recycling at local events, passing out flyers, holding essay contests at local schools, and running phone book recycling drives, the organization works hard to increase awareness and participation in community recycling. The coalition had had success in getting many citizens to voluntarily recycle, but the unavailability of curbside recycling kept recovery rates much lower than they could have been, and despite pressure from the coalition, the city had not been willing to include curbside recycling in its waste management plan. Then in 2011, the city got word that its current garbage dump was almost full and it would have to find a new trash service. The coalition saw an opportunity to push hard for a curbside program to be included in bids from potential new trash contractors, and thankfully, it was.

The final bid was presented to the city at a public meeting, which Lisa and other members of the coalition attended. At first, three of the five council members seemed to be against the plan. The mayor was concerned that adding recycling to the new trash pickup system might be too big of a change for the city. Lisa got worried. Nervous to speak up, but knowing that this was her one chance to voice the desires of so many people in the community who had been asking for years to have a curbside recycling program in place, Lisa stood up during the public comment period and spoke from her heart. She presented statistics she had researched before the meeting about how little Oklahoma actually recycles compared to how much the EPA says could be recycled and about how recycling could actually save the city money by diverting material from the landfill. At the end of Lisa's impassioned speech, the mayor smiled and said, "Well, I think you just convinced me." Lisa says that was the proudest moment of her life.

But the council still had to vote, and some members still seemed on the fence. When the votes came through with a unanimous "Yes," the room burst into applause, with the coalition members grinning from ear to ear. Lisa says that her years of serving on the board of the Ada Recycling Coalition and her one hour preparing facts before the meeting all paid off that night. "The feeling of being a part of change like this is something you can't describe. It made me so hopeful for the future and reminded me why I'm an activist."

In addition to her environmental activism, Lisa provides useful information on green living strategies on her personal blog, Retro Housewife Goes Green, and website, Green Oklahoma. She's a great example for people who live in rural areas without resources that might be more plentiful in bigger cities.

Drawbacks to Curbside Community Recycling Systems

While curbside recycling systems make it easier for residents to reduce the amount of trash we send to the landfill each year, there are serious drawbacks to depending on them for all of our recycling needs. A lot of disposable packaging waste is generated while we're away from home. Rather than lugging bottles, cans, wrappers, and take-out containers home for recycling, most people simply toss them in the nearest trash can. What's more, in many areas, curbside is only available to people who live in single-family homes. Apartment dwellers and office workers are out of luck. And while activist groups work to make curbside recycling accessible to more people, the truth is that communities must bear the cost of expanded recycling systems. In other words, you and I pay for it with our tax dollars. That may seem fair, since we're the ones generating the waste in the first place. But consider this: if product manufacturers were required to pick up all or part of the tab for recycling their packaging waste, they might be more inclined to reduce the amount of packaging they use in the first place or at least create packaging with recyclability in mind.

Extended Producer Responsibility Programs

Exended Producer Responsibility (EPR), otherwise known as Product Stewardship, is the philosophy that manufacturers should take responsibility for the full life cycle of the products they create. Within a product stewardship framework there are different ways to hold producers accountable, including take-back programs, in which companies directly take back and recycle their own products; fee-based initiatives, whereby manufacturers pay into the recycling system; and deposit programs like "bottle bills," in which consumers pay a refundable deposit on the goods they purchase and producers bear the costs of collection.

The forerunner of product stewardship schemes is the German Greet Dot system, which was implemented in 1991. Through that program, manufacturers are required to pay into the system for recovery and recycling or else take back recyclable packaging themselves. A green dot symbol on a product's package indicates to the consumer that the company has paid into the system. Fees are based on the materials used in packaging, so it encourages producers to use less. While U.S. states lag behind Europe in adopting EPR systems, examples do exist!

Product Take-Back Here are a few examples of successful voluntary North American take-back programs. Note that some of these companies downcycle returned items into secondary products that cannot be further recycled rather than completely closing the loop. But the point is that the companies are willing to bear the cost of processing rather than relying on community recycling programs.

- **Preserve's Gimme 5 Program** Preserve manufactures toothbrushes, razors, and other household products out of recycled #5 polypropylene (PP) plastic. Through its Gimme5 take-back program, customers can return their used Preserve products to Preserve to be recycled, along with yogurt tubs, food containers, or medicine bottles stamped with a #5, as well as products from select partners. Visit the Preserve Gimme5 website for the most up to date list of products accepted. Gimme5 drop-off bins are located at select food co-ops and many Whole Foods Market locations. Customers can also mail their #5 plastics back to Preserve.
- **HP's Print Cartridge Recycling Program** For years, HP has taken back its inkjet and laser printer cartridges for recycling into new HP products. Cartridges can be returned to participating retailers or mailed back to HP postage paid.
- **Tom's of Maine Recycling Program** Tom's of Maine has partnered with Terracycle in a program they call the Nature Care Brigade. You can send back empty toothpaste tubes, toothbrushes, floss containers, mouthwash bottles, soap packaging, and antiperspirant and deodorant containers for downcycling into things like duffle bags, garden pavers, and park benches. Visit the Tom's of Maine website for more information.

- **ChicoBag Bag Return Program** ChicoBag takes back all brands of reusable bags, including its own, to be made into new products or if still functional, distributed to low-income families for further use.
- **Patagonia's Common Threads Initiative** Not only does Patagonia take back and recycle used clothing into new clothing, it actively encourages customers to repair ripped clothing (themselves or through its repair program) and to look for secondhand clothing (through its partnership with eBay.com) before buying new Patagonia products. Amazing.
- **Brita Water Filter Cartridge Recycling** Through a partnership with Preserve, Brita has developed a take-back recycling program for its pitcher filter cartridges, which can now be returned via the Gimme5 program.
- **Dupont Tyvek Recycling Program** Many large envelopes, signs, banners, CD sleeves, and other products are made from Tyvek, a plastic product that feels like paper but doesn't rip like paper. Dupont has partnered with Waste Management Inc. to create a take-back program.
- **Dell's Global Recycling Programs.** Dell became the first electronics manufacturer to take back its products from consumers free of charge for recycling in 2006. The company offers a prepaid mail-back service, a drop-off program via participating Goodwill locations, asset recovery services for businesses, printer cartridge recycling, and more.

The bar has been set. Whether you like or approve of their products or not, companies like those listed above are leading the way in producer responsibility. When you see a new product advertised as "recyclable," why not contact the company and ask if they'll take it back?

Deposit Programs Currently, eleven U.S. states, all but one Canadian province, and many countries throughout the world have bottle deposit programs in place. Customers pay a deposit on the bottle when purchased and can get that deposit back when they return the bottle. Unrefunded deposits go into a fund to support the recycling program.

Producers must often pay into the program to support collection efforts. Studies show that states with bottle bills not only have much higher recycling rates (Bottle bill states recycle 48 percent of their plastic bottles, vs. 20 percent for non-bottle bill states),[84] but that the quality of the collected material is much higher than in states that do not have bottle bill legislation in place. And the programs are fair because only those who consume, produce, or sell bottled beverages bear the costs of recycling, not the general taxpayer.

Still, despite the benefits of bottle bills, the beverage industry as a whole has opposed them, believing that they inhibit sales. (Of course to me, the idea that bottle bills might curtail sales of bottled beverages is an added benefit.) The Container Recycling Institute has put together a website on current and proposed bottle bill legislation, as well as an activist toolkit for promoting bottle bills in your community. Visit www.bottlebill.org to get the facts and learn what you can do to promote container deposit laws in your community.

Electronics Take-Back Campaigns Electronic devices, such as computers, monitors, televisions, audio players and recorders, video cameras, and telephones, are not only a huge source of plastic pollution but of toxic chemicals and heavy metals like lead, cadmium, mercury, and brominated flame retardants. The EPA estimates that in 2012, the United States generated over 3.4 million *tons* of electronic waste. But only 29 percent of that was collected for recycling.[85] And unfortunately, just as most of our regular plastic recycling is shipped to developing nations, most of the e-waste that is collected ends up overseas in Asian nations and countries in Western Africa, where it poisons the people, land, air, and water. Children scavenge piles of trashed computer parts hunting for valuable materials, and villagers melt down plastics to recover any metals that can be resold.

Several organizations are working to put a halt to the practice of dumping electronic waste overseas and to require manufacturers to take responsibility for ethically recycling their products in an environmentally sound manner. Visit the website of the Electronics TakeBack Coalition to learn more about the problems of e-waste recycling and what you can do to hold manufacturers and recyclers accountable.

EPR Framework Legislation We may not be as far along as European countries, but some U.S. states are working on EPR framework legislation to hold producers accountable across the board. Instead of implementing separate laws for each type of product one by one (for example, bottled beverages or batteries or prescription drugs or fluorescent bulbs), an EPR framework is one law that establishes EPR as policy and gives the state government the authority, through regulation, to address multiple products. Framework legislation creates a level playing field among all producers.

UPSTREAM (formerly Product Policy Institute) works with local Product Stewardship Councils in the United States and Canada to develop framework legislation. Visit UPSTREAM's website to learn more about proposed EPR legislation or to join or start a Product Stewardship Council in your area.

Take Back the Filter: Anatomy of a Citizen Action Campaign

In the previous chapter, I described my dismay at finding out that my Brita water filter cartridge could be recycled in Europe but not in the United States. However, after learning about EPR programs such as the Green Dot system in Europe, I understood why. In Europe, producers are responsible for the full life cycle of the products they produce. In the United States, they are not. Still, we as their customers have the power to let them know what we want. I figured that if Brita filters could be recycled in Europe, there's no reason they couldn't be recycled here. So I wrote a letter to Brita and asked why. The company's response was unsatisfying. They sent me a form letter explaining that Brita recycling couldn't happen in the United States because community recycling systems were not equipped to handle water filter cartridges (and dismantling the cartridge and tossing the plastic housing in the recycle bin just makes its trip to the landfill a little longer). *Ha!* I said to myself. That's irrelevant. The Brita Company in Europe had built its own facility to recycle the filters. It wasn't relying on municipal recycling systems. And if Brita could do it there, surely the Clorox Company could do it here.

I ranted about my Brita filter recycling problem on my blog, and shortly thereafter, I discovered that the traffic to my website was increasing as people who wanted to recycle their Brita filters Googled "recycle Brita" and discovered my rant. Hundreds, maybe thousands, of other people wanted to recycle their filters! It wasn't just me. So in January 2008, I started a Yahoo! Group to discuss the issue and develop strategies for letting the company know what we wanted, and after a lot of hard work, by November of that same year, we got our wish. Brita had created a take-back program for its pitcher filters. Within those ten months, we had created an entirely Internet-based campaign, gathering over 16,000 petition signatures as well as 611 used Brita filters from all over the United States and Canada. Here's a rundown of the steps we took, which could be used as a model for other groups who want to hold companies accountable for their product waste.

1. Pick a Company That Gives a Darn. Clorox had already demonstrated its intention to "go green" by working with the Sierra Club to develop a line of green cleaning products, purchasing the Burt's Bees natural products line, and advertising its Brita filters as the environmentally friendly alternative to bottled water. The company was giving lip service to green ideals, so it seemed like it should be receptive to going further.

2. Pick an Issue You Care About. Staying motivated to keep a campaign going for ten months takes a lot of energy and persistence. Make sure the issue you're working on is something that's important to you personally in some way. I was already a Brita customer, so this issue affected me and my family. I cared.

3. Pick an Issue with Popular Support. Based on the hits I was getting on my blog, I knew other people cared about this issue. And based on the fact that Brita had developed a form letter to respond to questions about recycling, I knew the company must have received quite a few inquiries already. My job, it seemed, was simply to concentrate the power of our individual voices into one strong movement.

4. Do Your Homework. Before jumping into the campaign, I researched the recycling program that the German Brita company had created in Europe, as well as the recycling system in the United States, so I would know how to respond to Clorox's objections.

5. Plan Your Campaign with Other People Who Care. The Yahoo! Group was a meeting place for us to brainstorm and figure out our strategy. First, we all called and wrote letters to the company to see if we'd all get back the same form letter responses. In fact, we found the names of all the members of the executive committee in Clorox's Annual Report online and sent separate letters to each of them, asking to have a meeting. And we all got back the same form letters I'd received initially. We knew we would have to create a public campaign to get our voices heard.

6. Create a Simple, Direct Petition Asking for Specific Actions. There are lots of free or inexpensive petition websites out there. We used GoPetition.com, simply because it was cheap and easy to set up and use. Nowadays, we might use Change.org, Causes, or Care2. Then, we boiled our requests down to three simple statements, phrased in our terms:

> We, the undersigned, urge The Clorox Company to take responsibility for the millions of plastic filter cartridges that are landfilled or incinerated each year by:
>
> 1. Redesigning its Brita filter cartridges so that the plastic housing can be refilled rather than discarded each time the filter is changed.
> 2. Providing a take-back program, such as the one that exists in Europe, so that used cartridges can be returned to the company for recycling.
> 3. Creating a system for the cartridges to be dismantled and the components recycled/reused domestically rather than landfilled, incinerated, or shipped overseas.

As you can see, we were adamant that the company not simply pass off the plastic cartridges to recyclers in China, where they would just contribute to that country's pollution problems.

7. Create a Website with a Simple, Memorable Title. We named our campaign Take Back the Filter and used the free Blogger.com software to set up our website at www.takebackthefilter.org. Blogger is super easy to use. Even the most technically challenged person can create a blog or website in minutes. On the site, we included a link to the petition and background information about the campaign, and we posted news updates as the campaign progressed. We also included a running tally of filters collected.

8. Use a Visually Compelling Device. We asked supporters to mail us their used Brita cartridges as a way to make a strong visual statement. I rented a mailbox at the post office down the street so I wouldn't have to use my home address. Then, we waited. We had no idea if anyone would take the time to package up their filters, take them to the post office, and pay for shipping. But sure enough they did! Slowly, the filters starting coming in, arriving with more frequency as the months went by. Accountant that I am, I once again created a spreadsheet, this time to track the filters. All told, we collected 611 cartridges from thirty-nine U.S. states (including Alaska and Hawaii) and two Canadian provinces. Some people wrote messages and drew funny pictures on their filter cartridges before mailing.

9. Contact Everyone You Know. Once the website and petition were up, we set a launch date and contacted all our friends, other bloggers, representatives from environmental organizations, news sites, and anyone else we could think of to help spread the word. By that point, I had been blogging for several months and had developed relationships with many other bloggers who could help promote the campaign.

10. Use Social Media. Eventually, we also created a MySpace page, a Facebook page, and a Twitter account for the campaign. A local filmmaker created and donated an awesome YouTube video for us. Of course, social media sites come and go, and their popularity ebbs and flows. Use whatever social media platforms will be most likely to reach your target audience.

11. Make It Easy for People to Spread the Word. In addition to asking bloggers to write about the campaign, I also created a little square badge that they could post on their sidebars that would link to the campaign site. And I sent them sample posts with basic information in case they didn't have time to come up with an original post themselves.

12. Recognize Your Supporters. On one page of our website, we kept a running list of bloggers, organizations, businesses, and news sites that had written about the campaign or asked to sign on.

13. Keep Your Supporters Updated. As people signed the petition, I gathered their email addresses and sent them one email asking them to confirm if they'd like to be added to our mailing list. For those that replied, "Yes," I subscribed them to the blog news updates to keep them in the loop and keep momentum going.

14. Send Out Press Releases. For me, that was the hard part. I had no idea how to write a press release, so I had to learn. Fortunately, the Internet is full of helpful sites explaining how to do it; what format to use, and how to send it out. Eventually, we started to get some press, including a mention in the *NY Times*!

15. Stay Positive. Although some media sites wrote about the campaign in an Us vs. Them sort of way, I never felt like we were fighting against Clorox, but that we were simply helping the company to see what their customers wanted. Maybe even more importantly, I felt that our purpose was to educate the public on the issue of Producer Responsibility. No matter what the outcome of the campaign, I was sure we had at least done that.

16. Make It Fun. Invite people to your house to eat pizza and create a Brita water filter costume out of white poster board to wear across town during the Bay to Breakers race across San Francisco the next day. Or something like that.

17. Don't Give Up. I'll admit there were times during the campaign that I wondered what the heck I had gotten myself into. But I couldn't quit. Just as I had persisted in trying to find water at the Outside Lands Music Festival, I was not willing to back down from this pursuit. So many people and organizations stepped up to help out and spread the word along the way, I felt like I'd let everyone down if I threw in the towel.

In November of 2008, the brand manager from Brita called me personally to let me know that the company had partnered with Preserve to take back and recycle the pitcher filters through the Gimme5 program. The program didn't meet all of the criteria in our petition, but it came close. The filters would be recycled at Preserve's facility in the United States instead of overseas, and Brita would invest its own funds in getting the program going. Unfortunately, the program only included Brita's pitcher filters, which are made from #5 polypropylene, not its faucet filters, which are made from a different kind of plastic. And we would have preferred for the company to design a filter that could be opened and refilled instead of ground up and recycled into secondary products. But we'd come a long way from earlier in the year when Brita told us it couldn't recycle the filters because community recycling systems couldn't handle them. And frankly, at that point, I was exhausted and ready to move on to other things.

Take Back the Filter Brita Recycling Campaign

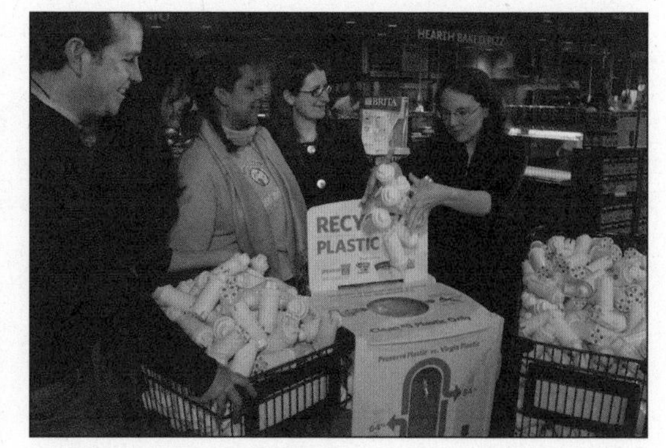

Is there a particular product you'd like to see taken back for recycling? Speak up. Start a campaign. It takes a lot of work, but the rewards are worth it.

Extended *Consumer* Responsibility

By promoting EPR programs, I don't mean to imply that the burden for dealing with our waste should be shouldered entirely by the product manufacturers. You and I have the choice to buy disposable products or not, and if we're going to buy this stuff and enjoy the convenience of it, we have a responsibility to make sure it is disposed of properly. If we buy a drink in a plastic bottle while out in the world, we can bring it home with us to recycle rather than throwing it in the trash. While travelling away from home, we can take steps to find out what recycling resources exist in the cities in which we find ourselves and either make some extra time to take our waste to a recycling facility at the end of the trip or make extra room in our suitcases to bring it back with us. And we can research other recycling resources for dealing with the stuff we can't put in our recycle bins.

Other Recycling Resources In addition to municipal recycling systems and producer take-back programs, there are other resources we can use to keep waste out of the landfill. Here's a list of just a few:

- **Terracycle** collects used juice boxes, food and snack wrappers, pens, markers, tape dispensers, containers, and many other household items, including some electronics, and downcycles them into secondary products. Honestly, I'm ambivalent about this program because while it's important to keep this crap out of the landfill, Terracycle does nothing to encourage producers to reduce the amount of packaging they create or to design for recycling. In fact, I think Terracycle lets companies off the hook. On the other hand, what are you going to do with that stack of dried up pens you've been hoarding?
- **CapsNCups** collects all bottle caps and #5 plastic cups to downcycle into an array of secondary products ranging from mud flaps to stadium seats.

• **Earthworks Gift Card Recycling Program** collects gift cards made from PVC plastic and recycles them into new gift cards.

• **Lions Club Eyeglasses Recycling Program** has bins at most retail eyeglass establishments and other locations for dropping off your old glasses to be donated to needy people.

• **Nike Reuse-a-shoe Program** collects used athletic shoes and downcycles them into flooring surfaces for gyms, tracks, and playgrounds.

• **Atayne Polyester Shirt Recycling Program** recycles any brand of fully worn 100 percent polyester performance tops into new polyester technical shirts and jerseys and provides a $5 credit to use at the Atayne online store.

• **The Bra Recyclers** collects used bras in good condition to distribute to needy women around the world. This service is not for worn-out bras but those that are simply unneeded, such as ill-fitting bras, maternity bras, or bras no longer necessary after breast surgery.

• **The e-Stewards Initiative** is a project of the Basel Action Network (BAN), which works to make sure that computers and other electronic devices are recycled responsibly and to prevent transportation of hazardous materials to third world countries. The site provides a listing of certified e-steward recyclers near you.

• **Green Disk Techno Trash Recycling Program** accepts CDs, DVDs, cassettes, floppy disks, hard drives and other media to refurbish or downcycle into other office products like jewel cases for new CDs and DVDs. The company also accepts used electronic devices but recommends customers first take advantage of local options if they exist. According to the company, no hazardous materials or obsolete components go overseas to be processed or disposed of.

• **Earth911** maintains a comprehensive directory for finding recyclers in your area. Type in your location and what kind of item you want to recycle, and Earth911 will provide a list of available options. The company also provides a toll-free, bilingual hotline, as well as a free mobile app.

Does Recycling Save Energy?

Advocates of recycling point out that the more recycled material we incorporate into new products, the less virgin material we have to produce/extract. A 2010 Life Cycle Inventory (LCI) of 100 percent postconsumer PET and HDPE found that the energy cost of making plastic bottles and then making polar fleece or other products out of them is 40 percent lower than it would be to make the bottles, throw them away, and then make that secondary product out of virgin resin. According to the study, it requires an initial 31.9 million Btu of energy to produce one thousand pounds of virgin PET pellets, and an additional 7.2 million Btu to recycle them into pellets again for a second use. That's a total of 39.1 Btu over the total life cycle of that resin, as opposed to the 63.8 Btu required to manufacture all virgin resin.[86] It sounds like a bargain, until we ask ourselves whether both "lives" were necessary to begin with.

Most of the PET and HDPE recovered for recycling comes from single-use disposable products—bottles, jugs, etc. The LCI doesn't take into account the energy needed to turn virgin PET pellets into single-use plastic bottles, fill them with water or other beverages, ship them to the store, and transport them home. And it doesn't show what happens when we remove the plastic bottles from the equation altogether. What if we didn't have disposable bottles to make our polar fleece from? Would we simply make it out of virgin resin, or could we, like Patagonia, recycle our worn out polar fleece into new polar fleece, in a truly closed loop system? What if plastics recyclers had to mine the racks of Goodwill for polyester to recycle instead having a constant supply of disposable bottles?

Every other week, I read about companies coming up with ingenious schemes for dealing with the plastic waste we generate. Recently, an Oregon start-up company called Agilyx developed a system for converting discarded plastic back into crude oil.[87] According to the *New York Times*, one factory module can turn 40,000 pounds of plastic into 130 barrels of oil per day, oil which can be converted into diesel, jet fuel, or other products. While these uses for our discarded plastic waste may seem like good ways to recapture the energy from disposable plastics instead of letting it waste away in the landfill for generations, consider the amount of oil we could have saved if we hadn't produced these disposable products and shipped them all over the country in the first place.

What if we stopped looking at recycling as a way to deal with the disposable packaging we use once and trash and instead as a way to increase the lifespan of the durable plastics we value?

Crafting and "Upcycling"

Rather than recycling plastic, which involves washing and grinding it down into pellets or flakes to make into completely new products, many people are finding other ways to repurpose used plastic packaging: cutting up soda bottles to make decorative lamps, weaving dresses from plastic bags, or stitching candy wrappers into wallets. Repurposing can be much less energy-intensive than recycling and, when done in a creative way, can be personally satisfying as well. It's also a good way to keep the plastic products we already have out of the landfill. A whole crafting movement has sprung up turning trash into treasures. Early in my plastic-free project, I myself knitted a "fake plastic fish" out of plastic bag yarn.

But there's a downside to the upcycling movement. As blogger Karen Lee wrote in her piece "Green Crafting: A Justifiable Means to an End?" on the website Crafting a Greener World:

> Is upcycling an excusable practice if we abuse or ignore the 'not-so-green' materials as a medium, all in the name of art and crafting? [. . . .] What if consumers feel complacent and are compelled to continue to buy plastic and not-so-green materials, thinking, 'Oh, I can make a planter with the bottle' or 'I can always make bags with this juice pouch so I don't feel guilty buying them'?[88]

It's one thing for artists and crafters to mine the dumpsters to salvage materials headed for the landfill. But as with recycling, does the existence of secondary uses encourage more consumption of disposable plastics?

On the other hand, some artists use recovered materials as a way to make a statement about waste and overconsumption. Their art can be a great educational tool if viewed as a call to reduce waste rather than as simply a way to deal with it. Here are a few plastics artists doing some amazing work:

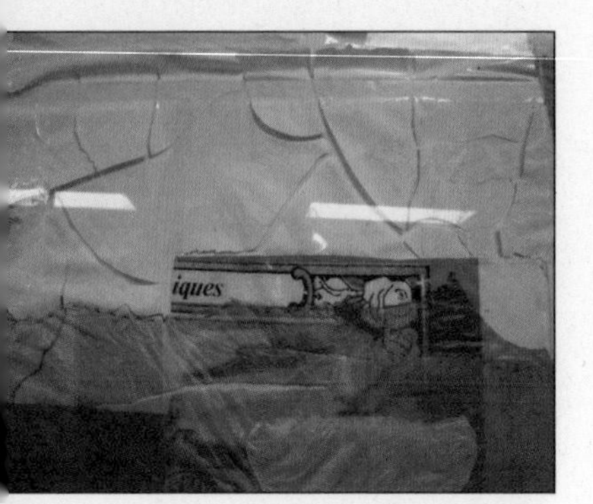

Dianna Cohen: Landscape. Plastic bags, handles and thread

- **Dianna Cohen**, one of the cofounders of the Plastic Pollution Coalition, is an L.A. artist who "paints" with plastic bags and other plastic waste instead of pigments. After working with plastic bags for about eight years, Dianna started to notice tiny cracks in her art pieces. Initially excited by this added dimension to her work, she was forced to rethink her notions of plastic, realizing that the disintegration of the plastic bags was part of the message about the flimsy nature of cheap plastic products ironically made of molecules that last forever in the environment .

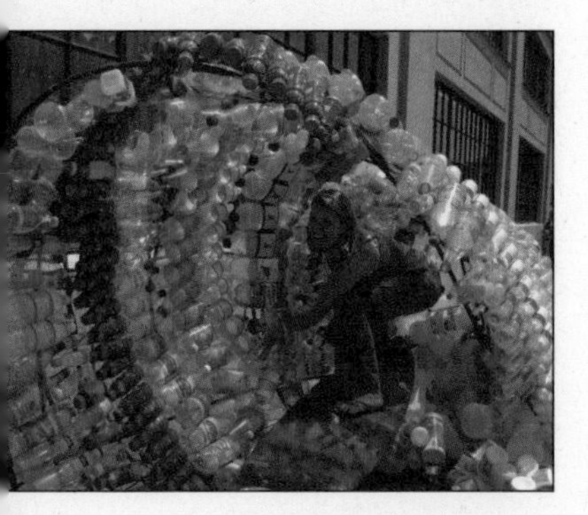

Kathleen Egan: Plastic Wave. Plastic bottles, steel, and chicken wire

- **Kathleen Egan** is a Bay Area surfer and head of the Plastics Subcommittee for the San Francisco Surfrider Foundation, an organization founded by surfers to protect oceans, waves, and beaches throughout the world. Kathleen creates mosaics and sculptures from plastic she collects from the world's beaches during surfing trips, and recently constructed a giant life-sized wave out of plastic bottles and other plastic trash donated by her friends. Visitors to the sculpture can climb inside the barrel of the wave and "ride" a real surfboard, surrounded by plastic "water." You can view a video of the plastic wave on YouTube. Search "The Plastic Wave" from YouTube user Mark Lukach.

- **Richard Lang and Judith Selby Lang** collect plastic trash that washes up on the beach at the Point Reyes National Sea Shore in Northern California and incorporate it into brightly colored collages and sculptures to highlight the environmental issues of plastic pollution. Visit their website at beachplastic.com.
- **Pam Longobardi** began collecting beach trash in 2006 after discovering the mountainous piles of plastic deposited by the ocean on remote shores of Hawaii. She calls her projects "interventions," as she moves thousands of pounds of plastic trash from the natural world to the art studio, where the objects can be examined in a cultural context.
- **Marina DeBris** uses humor and irony to show that the waste we create keeps coming back to haunt us. Creating dazzling costumes and other works of art from washed up beach trash, she encourages the viewer to question the use of single-use items and to consider ways to reduce waste so it does not end up in our oceans and landfill. View her work at washedup.us.
- **The International Plastic Quilt Project** asks participants to try living plastic-free for three months, collecting any plastic waste that they can't avoid, and to turn their collections into a quilt square at the end of the three months. The quilt is an amazing combination of colors, styles, and messages. The project is ongoing. Check the website to find out how you or your group can participate.
- **Tess Felix** "paints" portraits from the plastic trash she collects near her home at Stinson Beach, CA. From whimsical mermaids toting machine guns to expressive, intimate character studies, Tess's work elevates trash art to new heights. In 2013, she created a series titled "Ocean Eco Heroes," including portraits of Captain Charlie Moore, photographer Chris Jordan, activist Manuel Maqueda, and even me!

A great resource for discovering new trash artists and crafters is the blog EveryDayTrash.com. Personally, I love trash art. But I think it's important for us to view "upcycled" creations with a critical eye and ask ourselves what message the artist or crafter is sending. Are they making a statement about valuing the materials

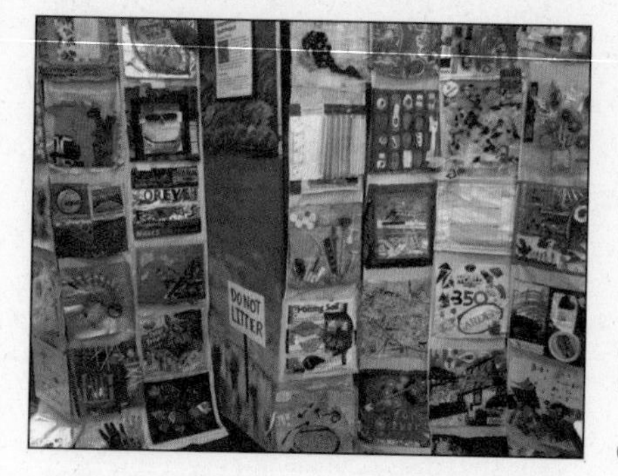

International Plastic Quilt

of the world and reducing waste and consumption? Or are they actually promoting consumption by creating tchotchkes that are destined for the landfill? As Karen Lee wrote me in an email, "There's 'Trash Art' and there's 'Trashy Art.' If 'Trash Art' makes a point, makes an argument against trash, then, it's serving its purpose. But 'Trashy Art' is just that—trash that took a short detour."

The difference between trash art and trashy art, of course, is a matter of personal judgment and individual taste, but it's important to ask the questions and to think before we buy.

Recycling Is Necessary, but It's Not the Final Answer

The reality is that there are many tons of disposable plastics already produced and that continue to be produced. Recycling slows down plastic's journey from cradle to grave, but ultimately plastics can only be recycled, downcycled, or upcycled so many times before they finally end up in the landfill or environment, where they will linger for thousands of years, or are incinerated or used for fuel, adding to air pollution and greenhouse gas emissions. And while companies are working hard to come up with innovative ways to recycle our plastic waste and organizations work to increase the availability of recycling facilities, the sad truth is that U.S. recycling rates are still abysmally low. In 2012, we recycled only 8.8 percent of the plastic waste we generated.[89] The rest of it ended up in landfills, incinerators, or the environment, while companies continued to produce more and more products from virgin plastic. Doing our part to recycle the plastic we use is important, but it's even more important for us as conscious consumers to "pre-cycle" by limiting the amount of stuff that has to be recycled in the first place. Let's follow the 4 Rs: **refuse** to buy single-use disposable packaging whenever possible; **reduce** the amount of unavoidable plastic we do consume and choose only those plastics that can realistically be recycled where we live; **reuse** plastic products when appropriate (some are not healthy to reuse); and **recycle** whatever is left. Finally, when purchasing durable products made from plastic, we can choose those made from recycled materials. In chapter 9, I'll discuss some useful products made from recycled plastic.

Action Items Checklist

(Choose the steps that feel right to you. Then, as an experiment, challenge yourself to do one thing that feels a little more difficult. Only you know what that one thing is.)

- ☐ If you have curbside recycling in your area, contact your recycling center and make sure you know exactly what items are accepted in the bin. Also, ask where the material goes when it leaves the facility. How much is shipped overseas and what products are created from it?
- ☐ If you don't have curbside recycling, research recycling centers in your area where you can bring your recyclables.
- ☐ Set up a recycling area in your home to separate materials and keep them clean and organized.
- ☐ Schedule a tour of your local recycling center to see the process for yourself. Bring kids if allowed.
- ☐ Arrange for recycling at your office. Organize a recycling station and volunteer to be the person in charge of making sure recyclables are sorted correctly.
- ☐ If you don't have curbside recycling, find out if there is an organization in your community pushing for it and get involved in the campaign. Or contact your city council and start a campaign yourself.
- ☐ Check the website www.bottlebill.org to find out the status of container deposit laws in your state and find out what you can do to get involved.

☐ Check the website upstreampolicy.org to learn about Extended Producer Responsibility legislation in your state and get involved in promoting it.

☐ Write to a company whose products you use and ask them to take back their packaging or used products for recycling.

☐ Join with others to start an action campaign to ask a company to take responsibility for its waste.

Chapter 5: Take-Out Food & Packing Lunches (Keep Your Straw, I Have My Own!)

Carrying our own bags and reusable bottles with us are great first steps for reducing our plastic waste. But so far, I've only addressed the first five items on my list! Staring at the piles of plastic I collected each week, I could see I had a lot of research to do to find plastic-free alternatives for all of it. But one category of items stood out as something I could reduce right away: take-out food packaging. While I might not know how to buy all of my groceries plastic-free, I could at least start making my lunches at home to eliminate the plastic clamshells, utensils, straws, soup containers, condiment packets and wrappers that come with take-out food. I brought my own bowl, plate, glass, and metal utensils to keep at work, so I wouldn't need to use my company's disposable foodware. And I started carrying reusable utensils and containers with me for the times when I did want to take restaurant food to go. Nowadays, there are many options for plastic-free reusable lunch containers, cloth snack baggies, sandwich wraps, and even drinking straws. The biggest challenge, I discovered, was paying attention.

Bringing your own foodware to a take-out restaurant is one thing. Remembering to speak up before the server wraps your sandwich in plastic or automatically sticks a straw in your drink is another. I found that I needed to be constantly alert to possible plastic that I'd have to bring home and add to my collection. It took practice, a lot of mindfulness, and the ability to laugh at myself and keep going when I forgot. Once, during a little road trip, my parents and I stopped at McDonald's to get a snack. I wanted an ice cream sundae. And I thought a sundae would be safe, since I had brought my reusable bamboo utensil set with me. "I don't need a spoon!" I proclaimed proudly to the server. A few minutes later, my face fell on the floor when she handed me my sundae in a plastic cup with a plastic lid. Duh! How could I have forgotten that the ice cream itself would be served in plastic?

Quick Tip:

Treat yourself to an ice cream cone. Ice cream cones require zero container or utensil waste. When I do want to bring some home, I get my ice cream hand packed in my own reusable container.

I also had to learn to speak up. Just as I'd practiced saying, "I don't need a bag," while handing cashiers my reusable one, Michael and I have had to muster the nerve to ask for our food to be packaged in our own containers rather than disposable packaging. Not all restaurants are willing to do it, but some are happy to use our containers. For example, the guys at my neighborhood corner café have no problem packing a sandwich in my reusable stainless steel sandwich container when I'm running late for work and have no time to make my own. When I'm sick, Michael runs down the street to our local Chinese restaurant and gets me hot and sour soup in our own reusable tiffin. (The only challenge is catching the staff before they put the tiffin, which has its own convenient carry handle, into a plastic bag.) When I ran out of time to make a dish for a potluck Thanksgiving dinner party a while back, I got the cooks at a nearby Italian restaurant to pack some butternut squash ravioli in my own glass casserole dish for me. I waited a while before confessing to my friends that I hadn't made it myself. And for my birthday one year, Michael convinced the staff at a local Japanese restaurant to arrange a huge order of sushi on one of their beautiful wooden trays, which he presented to me at my surprise party. He'd offered to pay the restaurant a deposit, which they refunded when he brought back the tray. The only thing he forgot about were the little plastic condiment cups they used for the ginger and wasabi. And the fake plastic grass. But this is how we learn.

Some restaurants are not as willing to use our containers. And some seem to change their policies depending on who is working that day. I find that a smile and a certain amount of persistence go a long way in getting people to comply. But more important than convincing restaurant staff to fill my to-go containers has been learning to prepare my own meals or making the time to sit down in cafés and restaurants that serve food and beverages in durable dishware. (Of course I still have to remember to bring my containers with me in case there are leftovers.) My life, like everyone else's, speeds along at a faster clip than I'd like it to sometimes. Planning time to breathe, look around, eat slowly, and relax is important for cultivating the mindful attitude that's so necessary for living without plastic. Ha! If only I had more time to do it.

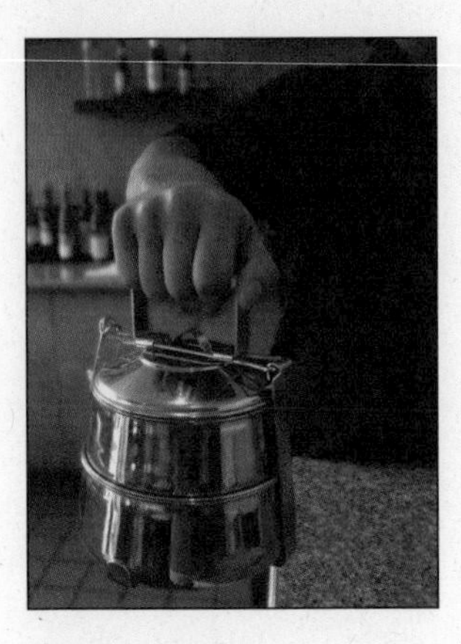

Picking up take-out restaurant food in my own To-Go Ware stainless steel tiffin

Whether or not take-out restaurants will use our own containers, we can all reduce the amount of restaurant packaging we do end up with by refusing the extras that we might not need. If we bring our own utensils, we can refuse the plastic ones. And we can refuse the extra ketchup, mustard, straws, lids, and wads of napkins (which, although they are not plastic, take a huge toll on the environment) that are automatically tossed into the bag. And of course, we can refuse the bag itself.

Take-out restaurant food in my own To-Go Ware stainless steel tiffin

The TakeOut WithOut Campaign provides downloadable wallet cards you can hand to servers when you are refusing packaging to educate restaurants on ways to reduce waste. And the Green Restaurant Association offers a directory of certified green restaurants that have committed to, among other things, reducing packaging waste. The Dine Green section of the site also provides downloadable wallet cards you can give to restaurants to encourage them to become certified.

Reusable Foodware for Lunches and Take-Out

Among the options for plastic-free reusable lunch containers are tiffins, cloth snack baggies, sandwich wraps, and even drinking straws. Of course, the greenest alternative is to use what you already have or to shop thrift stores or yard sales to find reusable lunch/food containers. But

"When my husband and I go out to eat, we split one meal so we don't have to use a carry-out container."
—Melanie Jade Rummel, Dallas, Texas

Quick Tip:

Instead of plastic packets of coffee creamer, ask restaurant servers if you can have plain milk instead. It will usually come in a small reusable pitcher or cup instead of disposable packaging.

Quick Tip:

When ordering pizza, tell the order taker you don't want the little plastic "table" in the middle of the pizza box. It's called a "package saver." Think about it: a single-use plastic device meant to save a single-use cardboard box. Tell them to skip the little plastic containers of parmesan and red pepper flakes, too.

from my experience, it's easier to find unwanted plastic Tupperware at Goodwill than sturdy metal or glass. The following list is not meant to promote consumerism but to provide plastic-free foodware options when repurposing or buying secondhand is not an option.

- **To-Go Ware** Stackable stainless steel lunch containers, small-sized snack containers, and bamboo utensil sets.
- **Life Without Plastic** Many different styles and sizes of stainless steel food containers, glass food containers, glass or stainless steel reusable straws, cloth lunch bags, insulated wool/cotton lunch bags, cotton/hemp sandwich and snack bags, and much more.
- **ECOlunchbox** Stainless steel bentos, lunch boxes, and artisan cloth lunch-bags and sacks.

- **LunchBots** Stainless steel sandwich containers, snack containers, and round insulated containers.
- **GlassDharma** Sturdy reusable glass drinking straws, guaranteed against breakage. GlassDharma invented the glass drinking straw.
- **PlanetBox** Stainless steel lunchbox with airtight compartments to hold an entire lunch in place. Comes with a cloth carrying case.
- **eco•ditty** Colorful, 100 percent organic cotton, plastic-free snack/sandwich bags.
- **Abeego** Waterproof cloth snack/sandwich wraps made from hemp/cotton infused with beeswax, tree resin, and jojoba oil. Abeego can be used to wrap all kinds of foods, as well as to cover bowls in the refrigerator instead of plastic wrap.
- **ReUseIt** All kinds of reusable containers, bags, utensils, straws, bottles, mugs, napkins, and a whole lot more. It's important to note that not all of ReUseIt's products are plastic-free. But the ReUseIt website allows customers to shop by material. Look for stainless steel, glass, organic cotton, hemp, ceramic, and other plastic-free materials.
- **Etsy** As I mentioned in chapter 2, Etsy.com provides a forum for individual crafters to sell handmade goods. There are many different sandwich/snack wraps and lunch bags offered through this site. But beware of ones made with PUL (polyurethane laminate), which has a plastic coating, or oilcloth, which is generally coated with PVC nowadays. Many of these fabrics have not been approved for food contact. Look for organic cotton, hemp, and other natural materials.
- **If You Care Greaseproof Paper Sandwich Baggies** While I am not a big fan of any disposable products, I do appreciate the care with which If You Care has designed its paper sandwich/snack baggies. They are made from unbleached, naturally greaseproof paper. I have to admit, I was skeptical about that claim. How could paper be greaseproof without some kind of coating? So I contacted the company to learn more. It turns out the barrier properties are

achieved by careful mechanical grinding of the pulp with no added chemicals or coatings. So if you need to use disposable baggies, this is probably your best bet. Read more about coatings on paper later in this chapter.

There are many other companies producing reusable sandwich/snack baggies and wraps that I have not mentioned in the above list because the products are coated with plastic or made from synthetic fabric. My preference is for plastic-free materials because I don't want plastic in contact with my food, but as I've said before, my choices are not necessarily right for everyone, and even synthetic or plastic-coated products can reduce a lot of disposable plastic waste.

"I keep a to-go container (I use a two-tiered tiffin) in the car so it's handy for leftovers. The tiffin is a perfect size for Chinese food to go. I've even crocheted a case for it, so it's not too hot to the touch, for either the server or me!" —Eve Stavros, Charlottesville, Virginia

PLASTIC-FREE HEROES:
Jay and Chantal from Life Without Plastic

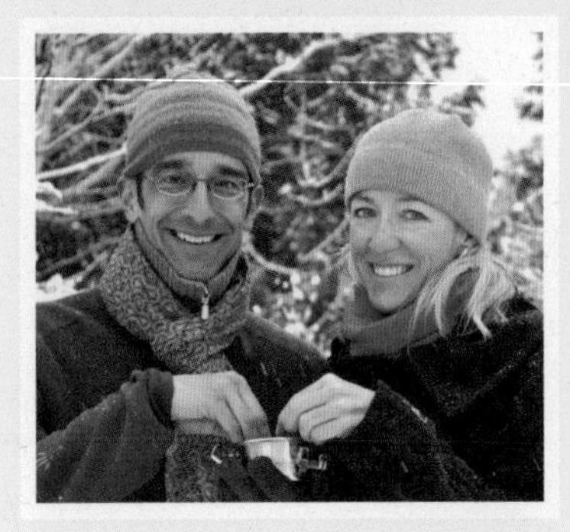

Canadian couple Jay Sinha and Chantal Plamondon had always been very health conscious and environmentally aware and had a feeling that plastic wasn't such a healthy material. They had noticed, for example, how tomato sauce would mix with the plastic in plastic storage containers, making them difficult to clean out. That couldn't be good. They switched to stainless steel water bottles when Klean Kanteens arrived on the scene. But before they had a baby, they hadn't done much other research into plastics.

After their son was born, Jay and Chantal began to look into the health impacts of plastic and decided that they didn't want their baby sucking on the stuff. They looked for glass baby bottles, but at that time, they could only order them directly from Evenflo in quantities of 1,000. So that's what they did, starting their company Life Without Plastic by selling glass baby bottles as well as Klean Kanteen water bottles.

After looking around for more plastic-free products and finding very few, and finding that what was available was often manufactured in China with low standards, they created their Sanctus Mundo brand—now known simply as Life Without Plastic—to design and manufacture their own line of stainless steel containers and other housewares with companies that treated their workers well and maintained high standards of quality. All of Life Without Plastic's branded products are ethically manufactured by trusted suppliers who meet the environmental and labor requirements laid out in their ethical sourcing policy.

Jay says it's been a challenge competing in the marketplace with other products that are less expensive because they are manufactured where labor costs are "ridiculously low." Jay and Chantal spend a lot of time researching companies and thoroughly testing their products before introducing them to the market, and they are extremely responsive to the feedback from their customers. I experienced this quality personally after Jay sent me a sample stainless steel container to review on my blog a few years ago. The container was plastic-free, but it came wrapped in a plastic bag inside the box. After I pointed out to Jay that the bag was a piece of unnecessary plastic waste, he immediately requested from his supplier that all future orders be shipped without plastic bags. And he has been very open to my ideas and suggestions ever since.

Jay and Chantal see the activist side of their business as just as important as the products they offer. Spreading educational information through their website, giving presentations at conferences, writing articles, and partnering with environmental organizations like the Plastic Pollution Coalition and 5Gyres, the couple hopes to create concentric waves of awareness everyday. Life Without Plastic regularly lends out water dispensers for events, and in their fair-trade certified home community of Wakefield, Quebec (Yes, towns can be certified fair trade!), Jay and Chantal helped put together a popular dish-lending library and spearheaded a plastic bag-free campaign.

I asked Jay what advice he would give to others who had a good idea for a business selling plastic-free products and weren't sure where to start. He said the first step would be to analyze your life. What interests you? What bugs you? What do you see around you that is lacking or that you would like to change? Base your business on what matters to you personally.

Jay also insists that there is no substitute for stellar customer service and sees Life Without Plastic's customers as teachers and guides from whom the couple is constantly learning.

Packing a Zero-Waste Lunch

Here are some ideas for packing lunches that leave no trash behind.

1. Start with a sturdy lunchbox, either new or used, or cloth lunch tote.
2. Main course: Sandwich in a stainless steel sandwich tin or cloth sandwich wrap or a hot meal such as stir-fry or mac and cheese in a stainless steel food container
3. Dry Snacks: Fruit, cut veggies, nuts, crackers, pretzels, chips, etc. in smaller stainless steel containers or cloth snack bags
4. Wet Snacks: Apple sauce, yogurt, chick peas, cheese cubes, etc. in small stainless steel containers or small reused glass jars.
5. Dessert: Pudding in stainless steel containers or cookies in cloth snack bags
6. Drink: Beverage of choice in reusable bottle
7. Cloth napkin
8. Stainless steel or bamboo utensils if necessary

Me and My Glass Straw

One of my favorite reusable items is my GlassDharma drinking straw. No, it's not necessary. In fact, many people question why we feel we need to suck up our beverages with a straw at all. That's a matter of personal preference. A straw can be helpful for people like me with sensitive teeth. And reusable straws made from glass are nice because they're see-through, so you can tell if they're clean or not. But if you're worried about breaking a glass straw (which has only happened to me once when I stupidly tried sticking a fork tine inside one to remove something stuck), there are stainless steel straws available as well. I keep my straw in a little cloth wrap, along with my bamboo utensils, but GlassDharma sells cases specifically for carrying their straws.

In addition to being fun to drink from, my glass straw is a conversation-starter and a way to get restaurant staff to pay attention when I ask for no straw. Many times I've requested no plastic straw when ordering a drink, only to have my drink delivered with a straw anyway. Wait staff are in a hurry and work on auto-pilot much of the time. By showing my reusable straw and explaining why I don't want a plastic one, I get their attention and help them remember not to automatically stick one in my glass.

"When I go out to a bar for a drink I always try to choose one of the beers on tap—practically packaging-free as I'm pretty sure breweries refill the kegs—plus you never get ambushed with a plastic straw!" —Joanna Morton, Wellington, New Zealand

One person I hope will remember my glass straw is *Project Runway* personality Tim Gunn. *Project Runway* is an addictive reality show in which fashion designers must complete all kinds of crazy challenges to win a fashion spread in a major magazine and big bucks to start their own line of clothing. Being highly critical of the fashion industry's role in promoting consumerism and waste doesn't stop me from marveling at the creativity of *Project Runway's* designer contestants who must work under strict time constraints to fashion clothing out of all kinds of unconventional materials, including scraps from a recycling center! So when I heard that the website BlogHer.com was sponsoring a contest for a lunch with Tim Gunn, I entered right away. I never thought I'd actually win. And when I found out I ***had*** won, I was nervous for days about what I would wear. Fortunately, I forgot my own self-consciousness for a second and thought about what I could bring Tim. I knew I didn't have to, but I wanted to bring him a gift to show my appreciation for the grace and generosity of spirit he displayed on the show each week. I decided to give him an elegant GlassDharma straw.

Meeting Tim Gunn at the Four Seasons in Chicago during the BlogHer conference is one of my favorite memories. Tim was as gracious in person as he seems to be on the show. He accepted my gift with apparent delight, exclaiming, "Oh, how novel," in his inimitable Tim Gunn style. Once again, the straw was a great conversation starter. Talking about reducing restaurant waste led to the topic of fashion based on reusing and repurposing old clothing rather than simply buying new stuff. Tim was receptive and easy to talk to. Plus, he told me I have a perfect hourglass figure. So there's that. The point is that even in a situation that ostensibly had nothing to do with environmental

ZERO-WASTE HEROES:
Milo and Odale Cress

Nine-year-old Milo Cress and his mom Odale used to write a food column together recommending healthy foods for kids. During the six months they wrote the column, Milo noticed a curious thing: a lot of diners would get their drinks and immediately take out the straw and lay it on the table. And then he noticed that even when he didn't ask for a straw, he would get one anyway. To Milo, it seemed like a huge waste of plastic straws. Why were servers putting straws in drinks if so many people didn't want them in the first place? And he wondered how much money restaurants could save if they only gave straws to people who wanted them.

So Milo and his mom did some research. They contacted the National Restaurant Association and found out that 500 million plastic straws are used in the United States *every day*. That's enough straws to fill up 46,400 school buses per year. Milo, who had been interested in environmental issues from a young age, started the Be Straw Free Campaign to encourage people to refuse plastic straws and to get restaurants to ask first before handing them out. His mom created a website where people can pledge to go straw-free for thirty days and restaurants can download a free table tent asking patrons to order their drinks "strawless." In February 2011, Milo testified before the Vermont Natural Resources and Energy Committee at the statehouse, and that same day he was able to meet the governor. From there, publicity for the Be Straw Free campaign took off. Milo has been featured on local television news shows as well as CNN, and the National Restaurant Association formally recognized his "Offer-First" suggestion as a best practice for restaurants nationwide. In the first eight months of the campaign, Be Straw Free gained support from individuals, businesses, schools, clubs and organizations in all fifty U.S. states, several territories, and over thirty countries around the globe. In fact, the Governor of Colorado named July 11, 2013, Straw Free Day to promote Milo's campaign.

Milo wants other people to help him spread the word to restaurants that are afraid to change their straw policies for fear of alienating their customers. He believes they will change if they know their customers want them to change, and he has outlined some steps to take to be the most effective:

1. Order your drinks with "no straw, please."
2. While you are eating at the restaurant, notice how drinks are served to other patrons. Do drinks have straws already in them or are straws offered when drinks are brought to the table?
3. If drinks are being served with straws automatically, ask to speak to the manager at the end of your meal after you've paid, and suggest that the restaurant ask before giving out straws. Let them know they can save money! If you notice that servers do ask first before giving out straws, thank the manager for this practice.

I asked Odale what advice she would give to other people who wanted to start a campaign. She said it's not always necessary to look before you leap because you can scare yourself off. But there are a few things you should think about: Is this project already being done? How streamlined and simple can I make it for people? And how can I keep it positive?

I asked Milo what he had learned so far from doing this project. He said, "I've learned that one person can make a difference. Sometimes we forget that every piece of plastic will be around for a long time. I want to live my life in a responsible way." As of this writing, Milo is on the second leg of an international speaking tour called "Let's Create the Future," to encourage kids everywhere to find and get involved in projects that interest them.

issues, I found a casual way to work the topic into the conversation. We don't all get to have lunch with celebrities, but we can find creative ways to bring up environmental topics with our friends and associates without nagging or preaching.

Zero-Waste Travel

Traveling away from home presents challenges to plastic-free living, but with a little planning, I've found ways to avoid generating plastic waste even while on the road or in the air. In fact, during the summer of 2010, I drove across the country with my dad and would have ended up with only a few plastic ice bags to add to my tally if Flamin' Hot Cheetos hadn't gotten the better of me. Here's how that particular incident went down:

Dad: [Returning to the car from a gas station minimart] I got us a treat!

Me: [Eyeing the plastic bag] Oh, I can't eat those.

Dad: Why? You don't like them?

Me: I *love* them and all their preservative-laden, artificially flavored and colored crunchy, spicy goodness!

Dad: [Smacking his head] Oh right! The plastic bag. Well, I guess I'll just have to eat all of them myself.

Me: Well . . . okay, just one . . .

What started with one Cheeto snuck from one bag turned into an orgy of Flamin' Hot Cheetos madness. My dad would buy a new bag of them every time we stopped for gas, and I'd munch out, until finally, towards the end of a long exhausting trip, I found myself buying the Cheetos myself. We were on a Cheetos bender that left me with a pile of plastic snack bags to add to my tally and the humbling realization that my devotion to reducing plastic use can sometimes be overcome by the temptation of laboratory-engineered, addictive junk.

I tell this story not to promote Frito Lay snack food but to reiterate that while we all do the best we can, none of us is a saint, and sometimes our humanity and the spirit of camaraderie can trump our best eco-intentions. The important thing is to be mindful of our choices. Flamin' Hot Cheetos notwithstanding, I brought home very little plastic waste from that trip. Here are a few of my traveling tips:

Bring Your Own Reusable Beverage Container—Even on a Plane! While it's true that transportation, and especially air travel, takes a huge toll on the environment, there's no reason to exacerbate that impact by giving in to disposable water bottles and food packaging. On a road trip, fill up your bottle at the soda stations at roadside minimarts and convenience stores. There's always a plain water option. When flying, bring your bottle through the security gates empty and fill it from the water fountain on the other side. I've never had a problem bringing an empty bottle through security or bringing a full bottle onto the plane. If you run out of water during a long flight, have the flight attendants pour drinks into your reusable bottle or cup. Yes, the water itself will most likely come from a plastic bottle, but you've at least saved a plastic cup or two. And during hotel stays, a reusable mug comes in handy when you find yourself in a room with only Styrofoam cups instead of real glasses.

Bring Your Own Snacks. Avoid plastic-packaged food. Bring your own sandwiches or containers of fruit, cut veggies, trail mix, or other snacks on the plane or on the road. Besides munching out on Cheetos during our trip, my dad and I snacked on nuts, trail mix, and other goodies I packed in the car.

Bring Your Own Utensils. Why should traveling be any different from staying at home? If we get in the habit of bringing our own utensils with us during short trips around town, bringing them during a long trip will be just as easy. Just make sure that if you're flying, you don't bring anything sharp. I've never had a problem bringing any of my bamboo utensils on a plane.

Refuse the Mini-Bar. Mini-bar drinks and snacks are incredibly expensive, and they all come in plastic packages or bottles. Find real food to eat. Do a little grocery shopping when you reach your destination and stock your hotel room with healthy snacks in less packaging. Even if you can't avoid plastic entirely, you can resist single-serving sizes.

Airplane snacks in my LunchBots stainless steel container

Bring Your Own Personal Care Products. Skip the free travel-size shampoos, soaps, and lotions provided by hotels. Just because they're free doesn't mean we should take them. They may not have a monetary cost, but they do have a cost to the environment and our health. Instead, fill up your own reusable travel-size containers at home. If you're flying, make sure your containers are small enough to comply with U.S. Transportation Security Authority (TSA) regulations or the airline security regulations in your country.

Bring your own utensils.

Reuse Your Ziploc Bag. When we are flying in the United States, the TSA requires us to bring all carry-on liquids in a 1-quart plastic zip-top bag. But there's no reason that bag should be tossed at the end of the trip. I've been using the same baggie over and over again for years.

Don't Forget Your Headphones. If you want to watch a movie or listen to music during your flight, don't forget to bring your own headphones. It can save money, since many airlines charge for headphones these days, and it will save plastic packaging and headphone waste.

Bring Your Own Pillow or Neck Rest. Airline pillows are not only made from synthetic fibers, but they are generally tossed out at the end of the flight! Oakland airport has led the way in developing a pillow recycling program, but the pillows are downcycled into insulation or furniture stuffing.[90] Bring your own and avoid the waste.

Collect Recyclables to Bring Home. I already mentioned this step in chapter 4, but it's worth repeating here. We might not be able to avoid all plastic packaging while on the road, but when we do generate recyclable plastic waste, we can either seek out recycling resources in the cities we visit, or we can do as backcountry hikers do and "pack it out."

Zero-Waste Parties

In December of 2010, after having worked at living plastic-free for three and a half years, for all the joy, creativity, and self-sufficiency I had experienced during that time, I was feeling a little burnt out and cranky. Maybe it was the winter blues. I had just come home from my friend Jen's beautiful holiday party, where I had spent much of the night worrying that even though Jen had provided real silverware, cloth bandana napkins, and other durable dishware, somehow plastic would enter the equation. I was tired of being on edge and feeling like a kill-joy some of the time. That night, I blogged sarcastically:

> Invite me to your party. I'll be the one who heads straight into your kitchen, opens the cupboard, and takes out a glass to use instead of plastic. I'll rummage through your drawers for reusable silverware. Or I'll take out the little bamboo set I carry in my purse. Your guests will find me charming.

I wrote a lot of other stuff that night about how ungracious I was feeling, and then went to bed and waited for the negative comments to come in. But for the most part, they didn't. My readers understood and commiserated with how challenging it can be sometimes to maintain happy relationships with family and friends and at the same time walk your environmental talk. Or even find ways to ***talk*** your talk without seeming naggy or self-righteous.

But the truth is that it hadn't actually been hard at all. My friends have been amazing, and while they, like most people, are not inclined to live as plastic-free as I, all of them have made big changes in their habits and perspectives in the last few years. As I mentioned, my friend Jen throws parties using reusable dishes and cloth bandana napkins; she carries a stainless steel water bottle,

> "I asked an Etsy seller to customize a purse for me with a lot of pockets that could hold a bottle, pen, straw, utensils, reusable bags, wallet, phone, and even a stainless steel food container . . . the purse holds it all and is not even very big or bulky!" —Melissa Brown, www.randomnessandfood.blogspot.com, Austin, Texas

for which she even made an ingenious little cozy out of an old sock; and she's found a myriad of other ways to cut waste in her home. My friend Mark religiously carries his reusable net bags with him when he goes grocery shopping. (And since the first edition of this book came out, he started sewing reusable cloth bags and selling them to his friends. Maybe by the time you read this, he'll have his own online store. If that happens, I'll be sure and list it on my website!) My sister Ellen called and asked me to help her figure out what recycling bins to get and who to call to set up curbside pickup. At his firm, Michael has taken on the role of "envirobusybody," explaining to new employees the three-bin recycling system in the building and reminding everyone what goes where. And for my part, I had simply been living my plastic-free life without nagging or lecturing the people I loved. If I was being ungracious towards anyone that night, it was myself.

But especially around the holidays, it can be challenging to speak up and suggest alternative ways of gift-giving or party-throwing that don't involve a lot of plastic waste. We don't want to insult our friends and family, and we worry about coming across as know-it-alls. But from my experience, I can say that a lot of that worry is simply in our own heads. It's the attitude with which we speak that can determine the reaction we get from people. For me, obsessing about offending people often turns into a self-fulfilling prophecy. But when I suggest an alternate way of doing things as an invitation for fun and creativity, the response is usually positive. Instead of saying, "Don't use plastic cups at the party because it will be bad for the earth," we can suggest, "Let's invite people to bring thrift store mugs and have a contest for the funniest slogan." We can always smile and add, "Plus, you know, the planet will thank us."

Plastic-Free July potluck at Berkeley Ecology Center

With these thoughts in mind, here is a list of zero-waste party ideas that might come in handy whether you're the host or simply a guest.

Bring Your Own Beverage Container and Utensils to Parties and Events. If you're not sure whether the host will offer real dishware or disposable plastic, discreetly bring your own. Or be less discreet, depending on your relationship with the host. I carry a little stainless steel wine glass (which is also good for events where glass is not allowed), which usually garners envy from other attendees who want to know how I lucked out and got "the good cup." People are intrigued when I tell them I brought it myself and usually want to know more. After all, most people would rather eat off of china, with silverware, than off of paper plates, with plastic utensils. I got mine from GreenBoatStuff.com, but you can also find them in camping supply stores.

"A couple friends and I gathered a box of items to use for parties. All of us had large BBQ parties this summer, but refused to use plastic. We went to all the thrift stores and bought silverware, enamel plates and bandanas (for napkins). This box now contains place settings for sixty-four people. Whoever uses the items washes them afterwards and keeps the box until the next person needs it. We also use pint canning jars for drinks such as water, ice tea, or lemonade." —Lori Tigner, proprietor, Westfarm Goats, Colorado

Stock Up on Thrift Store Utensils, Mugs, and Plates. You'll worry less about what happens to your dishes if you haven't spent a fortune on them to begin with. Plus, you'll keep perfectly good dishware from the landfill.

Ask Guests to BYO. Whether you ask them to bring a full set of dishes or just their own mugs or beverage containers is up to you. The more they bring and take back home with them, the less work for you.

Throw a Zero-Waste Potluck. Ask guests to bring dishes sans plastic wrap or disposable containers.

Provide Plastic-Free Beverages. Opt for homemade lemonade, iced tea, homemade soda, alcoholic beverages, or other store-bought beverages bottled in glass.

Skip the Goodie Bags. Back in the day when I was a kid, we took a present to the birthday boy or girl, and in return, we expected to play games, eat cake, and maybe wear a paper hat. We did not expect to come home with a plastic "goodie bag" full of plastic crap. Yet I'm told that goodie bags are the norm nowadays. Since I don't have kids, I asked my friend Danielle for advice. She told me that in her experience, kids don't care about goodie bags. As her seven-year-old daughter said, "You're supposed to go to parties to have fun!" A girl after my own heart. But if you do want to send kids home with a tangible memory of the party, why not involve them in some kind of craft—picture drawing, seed planting in paper pulp pots, cookie decorating—and let them bring home their own handmade creations?

"When I bring a dish to share to a friend's or a potluck, I cover it with a clean cotton bandana or large cloth napkin instead of plastic wrap. One bandana is placed over the top of the dish; another is brought up around the sides and tied firmly by diagonal corners Japanese 'furoshiki' style, which works nicely as a carrying handle too. I have several bright bandanas I keep for this purpose. I always get compliments about how nice it looks!"
—Maeve Murphy, San Rafael, California

Cities Saying No to Styrofoam

While there are all kinds of plastic foodware, "expanded polystyrene," or polystyrene foam ("Styrofoam" is a trademarked brand name), seems to be the most reviled. First of all, it's hard to recycle. According to the EPA, only 1 percent of all polystyrene waste was recycled in 2012.[91] Second, when littered, it crumbles apart easily and blows everywhere, making it very difficult to clean up. The wind carries it out to sea, where it can mimic food for marine animals. A California Department of Transportation study conducted during 1998-2000 found that polystyrene foam represents as much as 15 percent of the total volume of litter recovered from storm drains.[92] And finally, styrene, a suspected carcinogen, can leach from polystyrene food containers and contaminate our food.

While most municipalities are unwilling to ban all plastic foodware—or even the non-expanded version of polystyrene—altogether, many cities have moved to ban polystyrene foam. To date, 80 California communities have either banned polystyrene foam food containers completely or have prohibited use by government agencies or at public events.[93] Bans have also gone into effect in many other U.S. cities as well.

Check out Clean Water Action's "Phase Out Foam" page to see which cities have enacted ordinances. If you are in an area that has banned expanded polystyrene, and you notice that an establishment is still using it, politely remind the manager of the law. If you notice later that the eatery is still using it, you can notify the town's enforcement authority. Don't worry about driving the place out of business; the laws generally call for at least one warning before fines are levied. If the place is a franchise of a large chain, contact corporate headquarters and remind the company of its obligation to follow local ordinances.

Getting Styrofoam Out of Schools

At the risk of once again showing my age, let me just say that back when I was a kid, school lunches were served on durable dishes with metal utensils on trays that were washed and reused. Yes, the trays and dishes were all made from sturdy plastic, but they didn't become trash the minute lunch was over. Nowadays, many schools have ditched their dishwashing machines and instead serve kids their Tater Tots on disposable polystyrene foam trays and/or disposable plastic containers.

Some kids can simply avoid disposable plastic waste at school by bringing their own lunches. But the reality is that many kids rely on their school's subsidized lunch program to feed them healthy meals on nontoxic lunch ware. The quality of school lunches is a pressing issue beyond the scope of this book, but as for the lunch ware, there are campaigns afoot to get polystyrene and other disposable plastics out of our public school cafeterias. Here are some examples:

Portland, Oregon Portland Public Schools were going through 3.9 million polystyrene trays per year. So in 2007, with funding from Metro Regional Government to buy reusable trays, PPS started a permanent-ware tray pilot program to replace polystyrene trays in school cafeterias. Each school takes full responsibility for washing the durable trays, setting up all-volunteer washing programs with the help of parents, students, and community members. The schools have also switched to stainless steel utensils instead of plastic. How do they keep durable utensils from accidentally ending up in the trash? The Portland Public Schools website recommends keeping small buckets or containers at each table for the students to put their silverware in prior to getting up to dispose of their lunch trash. This prevents the silverware from ever making it to the trash can.[94]

New York City, New York According to the Styrofoam out of Schools campaign, NYC schools discard 850,000 polystyrene foam lunch trays ***every day***, which would make a stack eight and a half times as high as the Empire State Building. To combat the problem, NYC schools have instituted "Trayless Tuesdays." On those days, all 1,500 NYC schools serve "non-saucy" foods like sandwiches in paper boats instead of polystyrene foam trays. Any paper boats that are relatively clean can be recycled. To reduce plastic trash bag consumption, the campaign urges schools to "Flip, Tap, Stack" the paper boats to reduce the amount of space they take up. Students who are purchasing only one wrapped item and a drink are encouraged not to take any tray at all. And in schools that still have dishwashers, the campaign pushes for reusable trays. Visit the website for information on reducing Styrofoam waste in your school.[95]

Takoma Park, Maryland Since 2009, the Young Activist Club at Piney Branch Elementary School has been petitioning the school board for durable lunch trays and a tray washer. The club raised $11,000 towards the project and even hired a design consultant to conduct a feasibility study. The club feels it has enough money to start a pilot program, but as of this writing, the Board of Education remains unconvinced. Still, the students have accomplished a lot. They persuaded local businesses to sign a "no-styrofoam" pledge and to post window stickers. In June of 2010, the kids testified before the Takoma Park mayor and city council, which subsequently passed a resolution to ban use of city funds to purchase polystyrene food service ware and on

Nov. 10, 2014, passed the Young Activist Act of 2014, which bans polystyrene food-service ware citywide. And while the students might not have their dishwasher, they were able to convince the board to discontinue polystyrene foam trays and switch to paperboard based trays in the 2014-15 school year. They are considering using the funds they raised for the dishwasher to support a local composting project.[96]

Los Angeles, California After touring their local recycling center and learning that food-soiled PS foam does not get recycled, the kids from Thomas Starr King Middle School in Los Angeles decided to act. They collected 1,260 used foam trays from kids leaving the cafeteria (less than one day's worth!) and strung them together into a thirty foot art installation in the center of campus. They educated other students and petitioned the school board to eliminate foam trays. And their efforts paid off! In August 2012, The L.A. Unified School District (the second largest in the nation) officially banned polystyrene foam food trays from all campuses in the district.

Has your school system taken any steps to reduce plastic foodware waste? Does your school have a green team or club working on the issue? Check out the Plastic Pollution Coalition's Plastic-free Campuses initiative to learn more about what you can do to help your grade school or college make the switch.

Is Paper Foodware Greener?

As I mentioned in chapter 2, the production of paper has its own impact on the environment in terms of energy and water use and greenhouse gas emissions. Using durable foodware is a more sustainable option than simply switching to paper and cardboard. But there's another reason to avoid paper products that might surprise you: many of them are coated with plastic! That's right. Most paper products that are leak-proof and grease-proof are coated with some kind of plastic, even if it's hard to see. Examples include paper coffee cups, soda cups (the coating is not "wax"!), Chinese food cartons, paperboard boxes, paper soup containers, paper plates (except for plates that are not leak-proof), paper bowls, and while we haven't touched on grocery shopping yet, I'll just add paperboard

PLASTIC-FREE HERO:
Jordan Howard of Rise Above Plastics

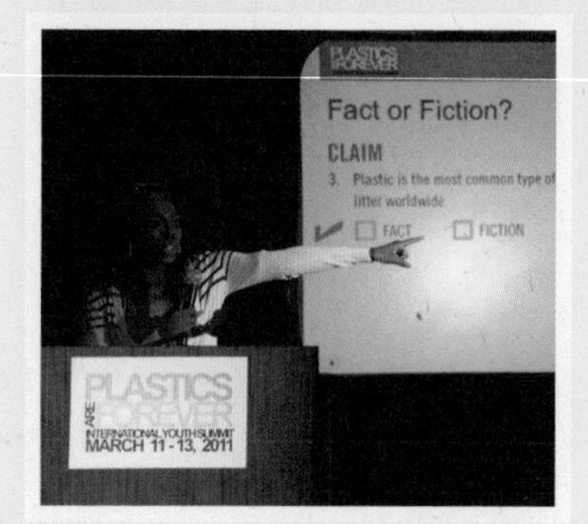

Jordan Howard wasn't always committed to protecting the environment. Even though she was a student at the Environmental Charter High School in Los Angeles, California—a school whose purpose was environmental education—she wasn't convinced that she should care about anything more than herself. A self-described typical teenage girl, she was interested in shopping and boys and whatever would benefit her in the short term. She says she was not "anti-sustainable," but she wasn't particularly pro-sustainability either.

And then, in her tenth-grade year of high school, Jordan was selected to attend the Bioneers environmental conference in San Rafael, California, not because she was an environmental activist but because her teachers recognized her as a leader. She went reluctantly, telling herself, ***I would not let those people brainwash me.*** But as she learned about all the different environmental solutions that had already been developed, she started to change her mind. "Compostable bags, recycled carpet, solar energy, organic clothing. I learned that living a green lifestyle was not only real, it was realistic, and most of the environmental and economic benefits were measurable!" That's when Jordan's natural leadership ability and her growing interest in green living started to merge.

Later that year, when Anna Cummins came to the school to give a presentation on plastics in the environment, Jordan was ready to listen and to take action. Plastic was something she used everyday, and she knew that her practical changes could make a difference. Bringing her own reusable bags, cup, fork, and using a stainless steel bottle were just some of the personal steps she took. But they weren't enough. Being a natural leader, Jordan wanted to encourage other kids her age to take action to prevent plastic pollution. So she teamed up with Surfrider Foundation's Rise Above Plastics campaign and the Green Ambassadors youth program to create the Rise Above Plastics (RAP) high school student speaker series. RAP not only educates high school teens about the harms of plastics on the environment and to humans, but empowers them to share the message and travel to schools and businesses across Los Angeles, educating people of all ages. RAP's goal for the first year was to give twenty-five presentations. They ended up giving fifty.

In 2011, Jordan worked with the Algalita Marine Research Foundation to organize the first annual International Plastics Are Forever Youth Summit, bringing together 100 students from fourteen different countries for two days of talks and training to find solutions to plastic pollution. (Visit AlgalitaYouthSummit.org for information on upcoming events.)

Jordan is now a college student, but she continues to work with RAP as a consultant and is thrilled that the program has been instituted as an actual class at Environmental Charter High.

I asked Jordan what advice she would give to other students who wanted to make a difference in their schools or communities but might not be sure how to start. Jordan says, "Start with small personal changes. As you perfect small things, educate other people. You will always want to do more and go bigger after perfecting the small things. Pass on the knowledge. And most important, have fun with it." You can follow Jordan Howard's latest projects at www.jordaninspires.com.

milk cartons, ice cream containers, cardboard frozen dinner trays, and any other paper or cardboard container that can hold wet food without leaking. These containers are generally coated with polyethylene. And of even more concern are the perfluorinated compounds I discussed in chapter 1 that coat many paper wrappers for burgers, sandwiches, microwave popcorn and even butter.

In municipalities where food-soiled paper and cardboard can be composted, the plastic coatings disrupt the composting process. Plastic coatings can make paper break down more slowly. What's more, plastic coatings themselves can break down into tiny micro-plastic particles that can contaminate the finished compost.[97] But we don't have to worry about possible environmental or health impacts of plastic-coated paper if we bring our own reusable containers and cloth wraps!

"To avoid perfluorinated compounds on fast food packaging, I take regular dinner fabric napkins to our local Subway Restaurant for the employees to wrap our sandwiches. I wash and re-use them many times."
—Jennifer Taggart, TheSmartMama, www.thesmartmama.com, Los Angeles, California

Bioplastics: A Better Alternative?

These days, we are seeing more and more plastics advertised as "Bio-based" or "Biodegradable," and some restaurants are switching to containers made from corn, sugar, bamboo, and other natural materials. Are these options better than plastic or plastic-coated paper? It depends. While there are innovative companies and scientists working hard to develop healthy alternatives to fossil-based plastics, some appear to be more interested in marketing their products than in making sure they actually do what they are supposed to. The messages can be confusing and the claims are not always verifiable. So I spoke with bioplastics expert Dr. Ramani Narayan from Michigan State University, who gave me a little primer. This is what I learned:

First, there's a difference between "bio-based" and "biodegradable" plastics. The term "bio-based" refers to what the plastic is made from: plants like corn, wheat, potatoes, sugarcane, or non-food plants, as opposed to fossil fuels. Bio-based plastics might or might not be biodegradable. The term "biodegradable" refers to what happens to products at the end of their lives: truly biodegradable products can be completely broken down by microorganisms in the disposal environment, which utilize the carbon in the plastic as food. Biodegradable plastics are not necessarily made from plants.

Bio-based plastics in general have a lower carbon footprint than fossil-based plastics because the carbon in them comes from renewable resources. Instead of releasing old carbon that has been sequestered under the ground for millions of years, manufacturers of these products use new carbon from plants that have absorbed as much carbon dioxide during their short growth period as they emit when they are disposed of. Keep in mind, though, that a lot of fossil fuels are used to grow and process the plants used to make bio-based plastics. For example, to produce plastics from conventionally grown corn, fossil fuels are used to power farm machinery, produce fertilizers and pesticides, transport crops to processing plants, and to process the raw materials. But Dr. Narayan says that despite all that, as long as the carbon footprint of processing bio-based plastic is the same or better than the carbon footprint of processing fossil-plastics, the plastics made from renewable, bio-based feedstocks have an advantage. Still, producers should continue to look for renewable alternatives to the fossil fuels used in processing.

Besides carbon footprint, there are other environmental impacts to consider when analyzing bio-based plastics. For example, while bioplastics ***can*** be manufactured from nonfood plants and plant waste, much of the bio-based plastic produced in the United States today comes from corn, and genetically modified corn (GMOs) at that, whose patents are owned by mega-corporations that threaten to monopolize the food supply. What's more, monoculture farming of crops like corn destroys biodiversity. The problems associated with industrial agriculture are beyond the scope of this book. Simply be aware that there are environmental impacts associated with whatever feedstock we use to create disposable products, and some are more problematic than others. Once again, reducing our consumption is key.

Okay, so now we understand "bio-based," right? Let's look at the term "biodegradable," about which there are a lot of confusing and misleading claims. Truly biodegradable plastics can be completely broken down by microbes in a particular environment. Knowing what environment they are meant to break down in is important because a product that is biodegradable in an industrial composting facility might not be biodegradable in a backyard compost bin or in the ocean, in which case it might be as much a threat to marine life as non-biodegradable plastic.

There are basically two kinds of plastic that are labeled (rightly or not) "biodegradable." Plastics that have a molecular structure that allows them to be processed completely by microbes are one kind. (Not all plant-based plastics have the right structure, as we'll see.) Plastics made from fossil sources that have an additive mixed in to force the plastic to break down are another. "Oxo-degradable" additives contain heavy metals to cause the plastic to start degrading after a specified amount of time. Some additives add starches in between the plastic molecules to weaken the plastic and cause it to break apart. And still others enhance the plastic molecules in some way, making them more digestible to hungry microbes.

But not all plastics labeled "biodegradable" or "compostable" actually break down completely in compost facilities. Recently, news reports have exposed the fact that some bioplastic utensils come through the composting process looking like new. And some supposedly biodegradable plastics really only degrade into micro-plastics, tiny pieces of plastic that pollute the soil and water. There is much controversy surrounding "biodegradable" claims. Some companies use dubious testing methods, such as analyzing what percentage of their product breaks down after a short period of time and using that information to extrapolate how much time it will take to break down completely. But Dr. Narayan says that these methods are faulty, and that the only tests that let us know for sure how well a material will break down in a particular environment after a particular amount of time are defined by established international (ISO) or national (ASTM, EN) standards (see notes for further explanation of the standards).[98] Fortunately for us, we don't have to understand all the ins and outs of test methods and standards because there are third-party certifiers that can verify the results and let us know which products are truly biodegradable, leaving no toxic residues, and under what circumstances. Any company making biodegradable claims should have their products certified by one of these organizations.

What you can do: No matter what the claims on the package, check to make sure any "biodegradable" plastic product you are considering buying bears one of these labels:

Third-Party Certifications for Bioplastics

<table>
<tr><td>COMPOSTABLE
Biodegradable Products Institute | US COMPOSTING COUNCIL</td><td>Biodegradable Products Institute (BPI) certifies products as compostable, which means they will biodegrade quickly, completely and safely, when composted in well-run municipal and commercial facilities and that no plastic residues will be left behind to destroy the value of the finished compost. This certification does not guarantee that the products will biodegrade completely in your backyard compost or that they will biodegrade in the ocean. You can check the directory on the BPI website to find brands of certified compostable bags, foodware, packaging, and even find out what companies have had their certification revoked.</td></tr>
<tr><td>®
compostable</td><td>European Bioplastics offers the "seedling" label, which certifies that the product will biodegrade completely in a well-run industrial compost facility, but not necessarily in a backyard compost or in the environment.</td></tr>
<tr><td colspan="2">Vinçotte is a Belgian organization that certifies products as compostable under the "seedling" label and also offers its own certifications:</td></tr>
<tr><td>OK biobased VINÇOTTE</td><td>OK biobased certifies the amount of bio-based material (carbon) in the product. Stars on the side of the logo indicate the percentage. The label does not tell you if the material is biodegradable or not.</td></tr>
<tr><td>OK compost VINÇOTTE</td><td>OK compost certifies the product will biodegrade completely in an industrial compost facility. However, it does not guarantee that the product will break down in a home compost system.</td></tr>
</table>

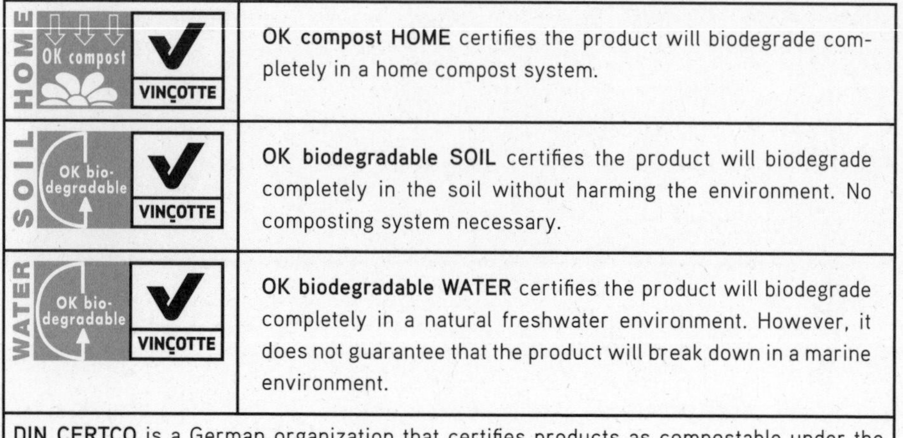

HOME OK compost VINÇOTTE	**OK compost HOME** certifies the product will biodegrade completely in a home compost system.
SOIL OK bio-degradable VINÇOTTE	**OK biodegradable SOIL** certifies the product will biodegrade completely in the soil without harming the environment. No composting system necessary.
WATER OK bio-degradable VINÇOTTE	**OK biodegradable WATER** certifies the product will biodegrade completely in a natural freshwater environment. However, it does not guarantee that the product will break down in a marine environment.
DIN CERTCO is a German organization that certifies products as compostable under the "seedling" label and also offers its own relevant certification:	
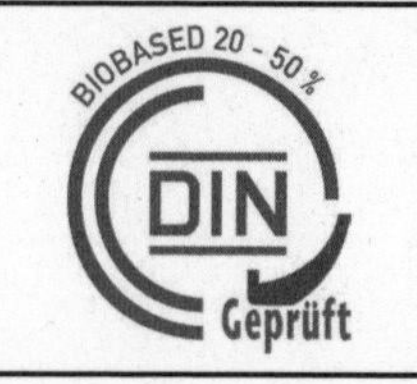	**DIN-Geprüft Biobased** This label indicates a product is certified to contain a certain percentage of bio-based material. The percentage is indicated on the label. This label does not tell you that the product is biodegradable.
Japan BioPlastics Association also offers two certifications.: **BiomassPla,** which certifies the amount of bio-based material in a product, and **GreenPla**, which guarantees a product will biodegrade completely in an industrial compost facility.	
Other certifications for biodegradability, compostability, and bio-based content may exist in other parts of the world.	

Common Bioplastic Products

Here are just a few of the many types of bioplastic you may encounter:

PepsiCo's Plant-Based Bottle and Coca Cola's PlantBottle are PET plastic made from plants. They are bio-based, but they are not biodegradable. For all practical purposes, they are just like regular PET plastic and can be recycled like PET.

Polylactic Acid (PLA) is bio-based plastic usually made from corn, although it could be made from any source of sugar. The primary producer of PLA in the United States is NatureWorks, LLC, an independent company invested in by Cargill and PTT Global Chemical, and its product goes by the brand name Ingeo. Ingeo is certified compostable by BPI, DIN CERTCO, and JBPA, which means it is guaranteed to biodegrade completely in an industrial composting facility, but not in home compost, plain soil, or the marine environment. However, you should check the label of any finished products made from PLA to make sure they are certified as well, since other materials could have been added, rendering the product non-biodegradable.

Polyhydroxyalkanoate (PHA) is bio-based plastic produced from sugar within the bodies of microbes. The primary company producing PHA is Metabolix, and its product goes by the brand name Mirel. To make Mirel, corn sugar is fed to specially engineered microbes that convert it to plastic through a process of fermentation. These microbes can produce up to 80 percent of their body weight in plastic. Mirel is certified compostable by BPI and Vinçotte. It is also certified home compostable, soil biodegradable, water biodegradable, and although there are no third-party certifiers for biodegradation in the marine environment or for anaerobic biodegradation, the U.S. Army Natick Soldier Research, Development and Engineering Center (NSRDEC) tested Mirel 5001 and found it to be fully marine biodegradable. And Organics Waste Systems (OWS) in Belgium found that it achieved 100 percent anaerobic biodegradation in 15 days. However, you should check the label of any finished products made from PHA to make sure they are certified as well, since other materials could have been added, rendering the product non-biodegradable.

Mater-Bi is a partially bio-based plastic that is made from substances obtained from plants, such as corn starch, and a combination of biodegradable polymers obtained both from renewable raw materials and fossil raw materials. Mater-Bi is manufactured by the Italian company Novamont and is used to make products like BioBag brand compostable bags. Mater-Bi is certified compostable by BPI, DIN CERTCO, and Vinçotte. It is also certified home compostable. However, you should check the label of any finished products made from Mater-Bi, since different products may not carry all of those certifications.

Bagasse is not actually plastic, but is sugarcane fiber that is pulped to make a kind of paper product, just as tree fibers are pulped to make paper. Bagasse can be molded into food containers, which, if uncoated, can be certified compostable. However, like paper and cardboard, some fiber-based products are coated with plastic, so check the label for compostable certifications. BPI lists several bagasse product manufacturers in its database, as well as manufacturers of molded pulp products from other types of natural fibers like wheat, bulrush, bamboo, and palm.

Degradable Additives Some plastics touted as biodegradable are merely fossil-based plastics combined with a degradable additive. Symphony's D2W and EPI's TDPA are oxo-biodegradable additives—they contain a heavy metal that causes the plastic to start breaking down mechanically. Earth Nurture Additive (ENA), ENSO Plastics, Bio-Tec's EcoPure, and ECM Biofilms's additives, on the other hand, are meant to attract enough microbes to the plastic to break it down. None of these additives has received third-party certification because none has proven it will biodegrade within the time frame specified by the standards. What's more, the Association of Postconsumer Plastic Recyclers is concerned that degradable additives may shorten the useful life of plastics and hinder their ability to be recycled.[99]

Keep in mind that compostable products are not meant to biodegrade in a landfill, where the goal is "mummification" rather than biodegradation. In the airless conditions of a landfill, bioplastics that do break down will release methane, a very potent greenhouse gas. While bio-based plastics generally have a lower carbon footprint than fossil-based plastics, we should do all we can to keep those that are truly biodegradable out of our trash bins. Compostable products are meant to be composted. To find an industrial compost facility that will accept bioplastics in your area, check out BioCycle's directory at www.findacomposter.com. But be aware that at this time, synthetic materials like bioplastics are not allowed in certified organic compost. It's important to understand what your compost facility will accept.

The website of the Sustainable Biomaterials Collaborative provides information addressing all of the possible environmental and health impacts of bioplastics and specifications and guidelines for choosing the most sustainable bio-plastic products. Remember: 1) Don't believe manufacturer's hype. Ask for the certification proving biodegradable claims, and 2) All materials, bio-based or fossil-based, have an environmental cost.

Bottom Line

Let's refuse disposable products of all types whenever possible and opt for bio-based alternatives when disposables can't be avoided. The point is to realize that we have choices and to do the best we can. After five years of attempting to live plastic-free, I still sometimes end up with unexpected straws, wrappers, condiment containers, or plastic-lined paper products. Let's cheer ourselves when we succeed, laugh at ourselves when we fail, and speak up to businesses and those who represent us to demand safer and less disposable packaging in the first place. We live in a throw-away society, and it's not going to change overnight. But each of our mindful choices and actions gets us one step closer to a healthier world.

Action Items Checklist

(Choose the steps that feel right to you. Then, as an experiment, challenge yourself to do one thing that feels a little more difficult. Only you know what that one thing is.)

- ☐ Obtain reusable foodware (new/used/or shop your own cupboards).
- ☐ Put together a reusable lunch/take-out kit.
- ☐ Bring reusable dishes, cup, and utensils to keep in the lunchroom at work.
- ☐ Make a plan for having your reusable foodware handy when needed.
- ☐ Talk to restaurant owners about reducing plastic take-out waste, offering reusables, giving a discount to customers who bring their own cups or containers, or switching to compostable foodware in areas where compost facilities exist.
- ☐ Contact local schools about lunchroom waste. Ask about switching to durable trays, utensils, and dishes.
- ☐ Organize your office to switch to reusable dishes, utensils, and cups instead of disposable.
- ☐ Participate in a cleanup of a local beach or waterway.
- ☐ Write a letter to the editor about the environmental and health impacts of polystyrene foam.
- ☐ Write to local and state legislators about bans on polystyrene and other disposable plastic take-out packaging.

Chapter 6: Grocery Shopping (Saving the Planet, One Cheese Wrapper at a Time)

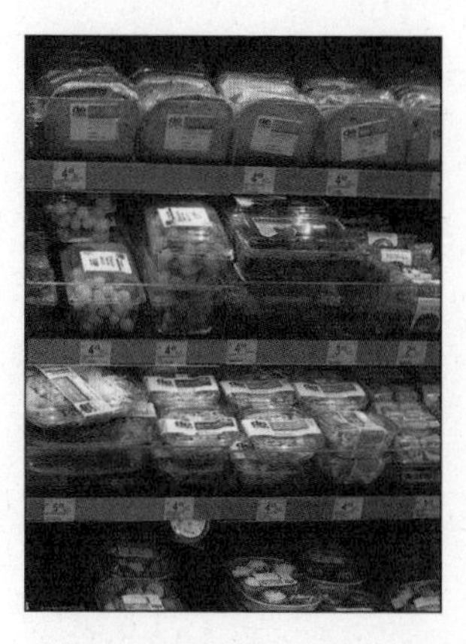

After finding ways to avoid bags, bottles, and disposable take-out and lunch containers, I was ready to get to the nitty-gritty: all my plastic-packaged groceries: pudding cups, hummus containers, grated parmesan cheese containers and lids, plastic juice bottles, frozen dinner trays and films, frozen vegetable bags, cheese wrappers, energy bar wrappers, candy wrappers, ice cream sandwich wrappers, chip bags, pasta bags, bread bags, raisin bags, soy milk containers and caps, the outside wrappers from boxes of tea and crackers, and the plastic bags inside cracker and cereal boxes. There was no way I was going to replace it all at once, and I didn't even know if I could. So, with notebook and pencil in hand, I set out on some fact-finding missions to learn what I could buy without plastic packaging. Like I mentioned, I was used to living on convenience foods, so strolling up and down the aisles of Whole Foods and Safeway, passing shelf after shelf of plastic

What plastic-free grocery shopping looks like.

packaging, I realized pretty quickly that my diet was going to have to change, in a big way. I might actually have to cook! After several weeks of research, I developed quite a few strategies for buying and preparing foods with less plastic.

Tips for Successful Grocery Shopping with Less Plastic

Say No to Singles

We've already said "no" to single-use disposable bags, bottles, and foodware. One step anyone can take to reduce grocery store waste is to buy foods in larger-sized packages and skip the single-serving sizes. I'm talking about those little containers of yogurt, pudding, and apple sauce, individual packets of nuts or chips, single-serve juice boxes, string cheese. I've even seen individually wrapped prunes and jellybeans! Buying larger sizes reduces the packaging to product ratio. We can portion out foods at home, packing yogurt or pudding for lunch in small reusable stainless steel containers or repurposed glass baby food jars. Or buying cheese in big blocks and cutting it into cubes ourselves. It might take a little more work, but it sure saves a lot of packaging. Plus, you get to decide how much you want at a time, rather than being bound by the manufacturer's idea of what a serving is.

Giving Up Packaged Frozen Foods

Here is how I learned to eat vegetables as a child: Go to freezer. Take out box of frozen broccoli. Open box. Dump broccoli into saucepan. Add water. Set pan on the stove. Turn on burner. Cook for five to eight minutes. Turn off burner. Drain broccoli in colander. Dump into bowl. Top with a glop of Cheese Whiz. Serve. Later, I would save time by using a microwave instead of the stove. And when companies started selling veggies with cheese sauce in microwaveable containers that let me cook and eat my food without even needing to put it in a bowl, I was thrilled. What could be more convenient?

The trouble is that, with few exceptions, all frozen food containers are either made of plastic or cardboard coated with plastic to keep them from leaking. One company, Stahlbush Island Farms, has switched to a "biodegradable" bag for its frozen produce. But the bag is lined with fossil-based plastic that contains a degradable additive to help it break down, and when I pressed a rep from the packaging company to tell me what chemicals were in the additive, he replied, "Ah, that's our secret ingredient." Without third-party certifications on the label to guarantee compostability or disclosure of the actual ingredients in the plastic, I for one am not inclined to trust these marketing claims. Still, this packaging is probably a step in the right direction, if only because it indicates that at least one processed food company has its heart in the right place.

After forty-some years of relying on frozen convenience foods for sustenance, I realized I was going to have to say goodbye to Green Giant, Stouffer's, Healthy Choice, and even natural brands like Amy's Kitchen, Organic Bistro, and CedarLane. And I was going to have to explore heretofore mysterious sections of the grocery store, like the produce aisle.

Produce wrapped in plastic.

Plastic-Free Fruits and Vegetables

The majority of the fruits and vegetables available from grocery stores in my area are packaged with little more than a plastic sticker or tag. At first, this seemed like the best I could do. After all, one plastic produce sticker is nothing compared to the box or bag used for frozen veggies. Later, I learned to avoid even those plastics. Here are a few tips for buying produce without plastic. Keep in mind as you read that it has taken me years to perfect my strategies and that you don't need to completely eliminate plastic packaging overnight.

Choose "Naked" Veggies. The produce aisle is full of choices, some of which contain more plastic than others. Avoid plastic-wrapped Styrofoam trays, plastic containers of cut veggies, baby carrots in plastic bags. Buy as many vegetables as you can without packaging, and for the rest, ask yourself if you can find them elsewhere or even

PLASTIC-FREE HERO: Kippy Miller from Kippy's Ice Cream Shop

If you live in the Los Angeles area, you might find one frozen dessert packaged in glass jars instead of plastic or plastic-lined cardboard: the coconut-based non-dairy ice cream from Kippy's Ice Cream Shop. Frozen foods in glass? Doesn't glass crack? "Not if the ice cream is processed correctly before it goes into the jar," says company owner Kippy Miller. She explained to me that when you freeze water in glass, the glass will crack due to expansion. The same thing would happen if she were to pour her liquid ice cream mix directly into glass jars. Instead, she first churns the mix in a big gelato machine that partially freezes it as it turns, allowing the ice cream to expand. Then, she can fill the glass jars with the already expanded ice cream, and they will not crack in the blast freezer. The method works. If it didn't, the brand wouldn't have lasted very long at the Whole Foods Market stores and other L.A. area outlets that carried it.

Since the first edition of this book, Kippy has closed her wholesale operations to focus on opening retail shops. At Kippy's Ice Cream shops, you can eat your organic, non-dairy ice cream on site in beautiful reusable glass or ceramic bowls with real silverware, or there are three ways you can take it to go. You can purchase a pre-packed returnable glass jar of ice cream and bring your jar back for a $.25 refund, take it out in a plastic-free bagasse container, or bring your own container to be filled at the shop. Just want a taste? Kippy uses real metal tasting spoons, not wood or plastic. She told me she loves talking to customers about plastic and has plans to open more shops, both in L.A. and San Francisco, in the coming years.

Kippy doesn't just package her product sustainably, but she expects her suppliers to deliver her ingredients without plastic as well. She refuses to use frozen fruit because it all comes packaged in plastic, opting for fresh organic fruit that can be purchased "naked." But, she laments, as her company grows and she has to rely more on produce distributors than individual farmers, it's becoming harder to avoid plastic packaging. Berries, for example, often come in plastic clamshells that the distributors will not take back to reuse. Still, she pushes for a plastic-free supply chain.

I asked Kippy how she got started in this business and why using plastic-free packaging was important to her. She told me that she hadn't always been in the food business but actually went school for fashion design and spent fifteen years in the fashion world, which involved a lot of traveling and eating out. The lifestyle took a toll on her body. And she started to notice how wasteful the fashion business was: disposable makeup containers; throw-away clothing; a culture of consumerism. She wanted to opt out of that world and become an integral part of her community. She met her future partner, Max, who made fresh ceviche each weekend and started teaching her about healthy eating and preparing raw foods, and she fell in love with coconut cream.

One day, after mixing up a concoction of coconut, honey, and strawberries, Kippy had the sudden realization: ***Oh, my God, I think this is ice cream!*** Excited, she started experimenting with different recipes and combinations until she had the perfect formula. She says that from the beginning she knew she would find a way to package her product in glass to avoid the chemicals that can leach from plastic. At first, people didn't believe she could do it. But Kippy advises anyone who has a unique idea, "Just because someone's never done it before, don't believe it can't be done."

if you really need them or if you can substitute a different, unpackaged fruit or veggie. You might decide you don't want to make a substitution, but it's important to ask the question.

Say No to Plastic Produce Bags. They are generally unnecessary. What are we worried about? That our apples won't get along with our broccoli during the trip home? Or is it that the produce will get dirty? My philosophy is that it grew in the dirt and has already been touched by

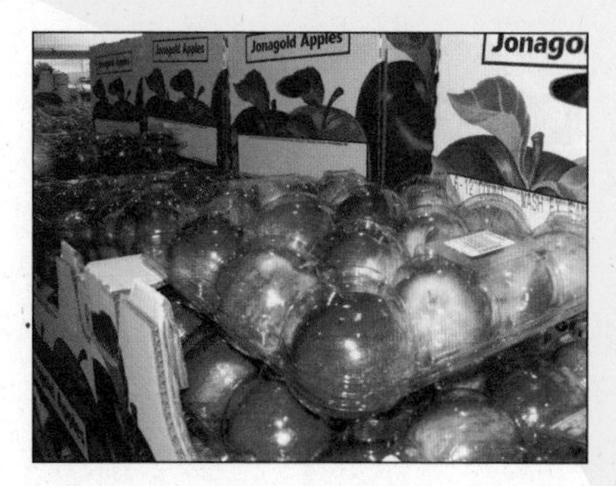

Produce packaged in plastic.

untold numbers of hands before it got to my shopping cart. Putting it in a produce bag is not going to make a difference between the store and my home, where I am going to wash it anyway. I put large produce items directly into the cart and later into my reusable shopping bag, and I use a cloth produce bag for smaller items. (Eco-Bags Products, ChicoBag, ReUseIt, and Etsy offer a wide variety of reusable produce bags, and natural foods stores often sell them right next to the produce these days.)

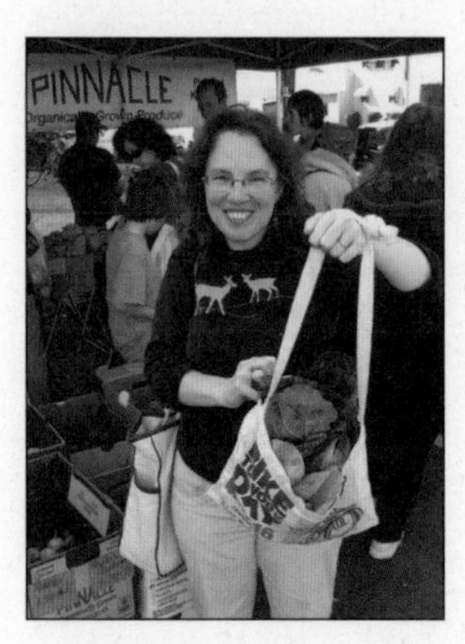

Buy naked veggies without produce bags.

"I bring old potato nets with me to the store and pack fruits and vegetables in them—they are very lightweight, so even if the store won't let me subtract their weight from the total, it's not a big issue."
—Rebecca Knecht, Freiburg, Germany

Shop Your Local Farmers Market. Farmers markets are a great way to buy fresh, local produce without plastic, as long as you remember to bring your own bags. And normally, the fruits and vegetables at farmers markets don't even have those little plastic stickers on them. Local Harvest provides an online directory of farmers markets in the United States and Canada. Some, like the ones in the Bay Area, are open year-round. Others, in areas with more extreme climates, are only open for part of the year. Why not explore your local farmers market and find out what kinds of produce options are available? One strategy is to first shop the farmers market each week, and then use the grocery store for whatever you still need to buy. You'll not only avoid plastic, but you'll reduce the energy needed to ship produce from far away and support local agriculture and small farmers.

Join a CSA. CSA stands for "community supported agriculture." It's a way for customers to buy directly from small farmers, either by visiting the farms themselves or having fresh produce delivered. If you do opt for delivery, make sure you let the CSA know that you don't want your produce packaged in plastic, as some of them use plastic bags. The Local Harvest directory lists CSAs in your area as well.

Return Containers for Berries, Cherry Tomatoes, Etc. to the Farmers Market to Be Reused. Most grocery stores won't take back plastic containers, since the produce is delivered by distribution companies that don't have the infrastructure to get the containers back to the producers. But the vendors at farmers markets love to have their containers back to reuse. Many vendors will even take back containers that came from the grocery store or other source. It saves them money and reduces waste. You can either bring your container back to the market on your next trip, or simply dump your berries into a container you've brought with you and hand back the plastic one to the vendor immediately.

Eat It Whole or Juice It Yourself. To avoid plastic juice bottles, consider buying whole fruit and juicing it yourself. For citrus juice like orange, lemon and lime, you don't even need a fancy electric juice machine. A manual juice squeezer or press works well. You can find simple models made from all stainless steel, ceramic, glass, and stoneware. Fresh-squeezed juice tastes better anyway. Or skip the extra work and eat the fruit whole. While there are juices bottled in glass, buying whole fruit generates less waste. And to me, fruit juice is simply concentrated sugar with most of the healthy fiber extracted, anyway. Instead of bottled lemon or lime juice, I keep cut wedges of lemon and lime in the freezer in glass jars. But in a pinch, you can find small glasses of lemon or lime juice. Lakewood and Santa Cruz Organic are two brands.

Grow Some Yourself. In 2011, I planted a veggie garden in my tiny front yard. To keep it plastic-free, I used flattened cardboard boxes instead of plastic ground covers; had bulk garden soil delivered instead of buying plastic bags of the stuff; planted seeds from paper packets into cardboard toilet paper tubes; purchased plants in compostable

My plastic-free gardening adventure

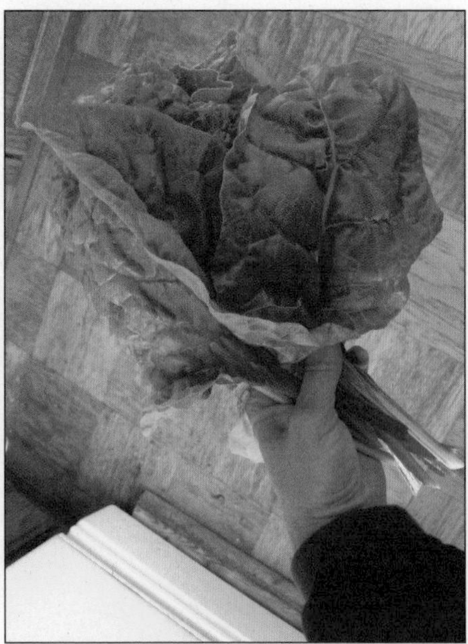

pots, or returned plastic pots to the vendors for reuse; and used a big metal watering can instead of a plastic garden hose. I threw myself into my plastic-free gardening adventure with as much enthusiasm as I had when I first started my plastic-free life. To be sure, there were trials and tribulations. My cucumbers never grew. My wax beans grew to about three inches and then turned brown and died. My squashes started out promising and then a few months later succumbed to powdery mildew and pollination problems. I battled aphids and leaf miners and learned the sex parts of zucchini. And all the while I kept in mind the admonition of Eric Hurlock from OrganicGardening.com to keep my hopes high and my expectations low.

Farmers market produce without stickers

The day I picked my first tender chard leaves, brought them upstairs to my kitchen, sautéed them immediately along with some of my own fresh-picked oregano, and served them to myself on a bed of bowtie pasta, I wept. I'm not kidding. The feeling of pride and self-sufficiency that came over me was indescribable. *I could feed myself*. And call it my imagination, but I swear those veggies tasted miles better than any from the grocery store or even the farmers market. Whether you have room for a full-fledged garden or only a couple of terra cotta pots, try growing something edible. You might be amazed at what you discover about yourself and your connection to the planet.

Interesting Fact:

Ever wonder why organic bananas have plastic around the stems but non-organic bananas do not? The reason is that nonorganic bananas are dipped in fungicide to prevent mold. On organic banana stems, plastic is used instead.

Avoid Produce Stickers When Possible. Plastic produce stickers can be annoying, but they serve a purpose in grocery stores. First, they tell us the country and often state of origin of the food, which is important if you want to reduce your food miles. Second, the numeric PUL codes on them help us and grocery store employees differentiate between organic and nonorganic foods. A four number code means nonorganic. A five-number code beginning with 9 indicates organic food. But while the stickers can be useful, they also make trouble for waste water treatment facilities, where they can get stuck on pumps and hoses, caught in screens and filters, or even worse, make their way into our waterways. And they are trouble in compost as well. So always remove produce stickers before peeling fruits and vegetables. Do not let them go down the drain, garbage disposal, or into your compost. What should you do with them? If you don't want to throw them in the garbage, you can send them to Barry Snyder of Stickerman Produce Art (his mailing address is on his website), who makes amazing collages with them. Or keep them and make collages yourself. And of course, you can avoid produce stickers entirely by shopping the farmers market.

QUICK LIST: HOW TO STORE FRUITS AND VEGETABLES

Tips and Tricks to Extend the Life of Your Produce without Plastic

Reprinted with permission of the Berkeley Ecology Center, which runs the Berkeley Farmers Market (www.ecologycenter.org).

Fruit:

Apples—store on a cool counter or shelf for up to two weeks. For longer storage, put in a cardboard box in the fridge.

Citrus—store in a cool place, with good airflow, never in an airtight container.

Apricots—store on a cool counter to room temperature or fridge if fully ripe

Cherries—store in an airtight container. Don't wash cherries until ready to eat, any added moisture encourages mold.

Berries—don't forget, they're fragile. When storing be careful not to stack too many high, a single layer if possible. A paper bag works well, only wash before you plan on eating them.

Dates—dryer dates (like Deglet Noor) are fine stored out on the counter in a bowl or the paper bag they were bought in. Moist dates (like Medjool) need a bit of refrigeration if they're going to be stored over a week, either in cloth or a paper bag—as long as it's porous to keeping the moisture away from the skin of the dates.

Figs—don't like humidity, so, no closed containers. A paper bag works to absorb excess moisture, but a plate works best in the fridge up to a week unstacked.

Melons—uncut in a cool dry place, out of the sun up to a couple weeks. Cut melons should be in the fridge, an open container is fine.

Nectarines—(similar to apricots) store in the fridge is okay if ripe, but best taken out a day or two before you plan on eating them so they soften to room temperature.

Peaches—(and most stone fruit) refrigerate only when fully ripe. More firm fruit will ripen on the counter.

Pears—will keep for a few weeks on a cool counter, but fine in a paper bag. To hasten the ripening put an apple in with them.

Persimmon—Fuyu (shorter/pumpkin shaped): store at room temperature.

Hachiya—(longer/pointed end): room temperature until completely mushy. The astringentness of them only subsides when they are completely ripe. To hasten the ripening process place in a paper bag with a few apples for a week, check now and then, but don't stack—they get very fragile when really ripe.

Pomegranates—keep up to a month stored on a cool counter.

Strawberries—don't like to be wet. Do best in a paper bag in the fridge for up to a week. Check the bag for moisture every other day.

Veggies:

Always remove any tight bands from your vegetables or at least loosen them to allow them to breath

Artichokes—place in an airtight container sealed, with light moisture.

Asparagus—place them loosely in a glass or bowl upright with water at room temperature. Will keep for a week outside the fridge.

Avocados—place in a paper bag at room temp. To speed up their ripening, place an apple in the bag with them.

Arugula—arugula, like lettuce, should not stay wet! Dunk in cold water and spin or lay flat to dry. Place dry arugula in an open container, wrapped with a dry towel to absorb any extra moisture.

Basil—is difficult to store well. Basil does not like the cold, or to be wet for that matter. The best method here is an airtight container/jar loosely packed with a small damp piece of paper inside—left out on a cool counter.

Beans, shelling—open container in the fridge, eat ASAP. Some recommend freezing them if not going to eat right away

Beets—cut the tops off to keep beets firm, (be sure to keep the greens!) leaving any top on root vegetables draws moisture from the root, making them lose flavor and firmness. Beets should be washed and kept in an open container with a wet towel on top.

Beet greens—place in an airtight container with a little moisture.

Broccoli—place in an open container in the fridge or wrap in a damp towel before placing in the fridge.

Broccoli Rabe—can be left in an open container in the crisper, but best used as soon as possible.

Brussels Sprouts—if bought on the stalk leave them on that stalk. Put the stalk in the fridge or leave it in a cold place. If they're bought loose, store them in an open container with a damp towel on top.

Cabbage—left out on a cool counter is fine up to a week, in the crisper otherwise. Peel off outer leaves if they start to wilt. Cabbage might begin to lose its moisture after a week, so, best used as soon as possible.

Carrots—cut the tops off to keep them fresh longer. Place them in closed container with plenty of moisture, either wrapped in a damp towel or dunk them in cold water every couple of days if they're stored that long. [Beth's note: I cut up carrots into pieces and store immersed in a container of water in the refrigerator. They stay crisp for many days.]

Cauliflower—will last a while in a closed container in the fridge, but they say cauliflower has the best flavor the day it's bought.

Celery—does best when simply placed in a cup or bowl of shallow water on the counter. [Beth's note: As with carrots, I cut up stalks of celery and keep immersed in a container of water in the refrigerator for snacking. The water keeps them nice and crisp.]

Celery root/Celeriac—wrap the root in a damp towel and place in the crisper.

Corn—leave unhusked in an open container if you must, but corn really is best the day it's picked.

Cucumber—wrapped in a moist towel in the fridge. If you're planning on eating them within a day or two after buying them they should be fine left out in a cool room.

Eggplant—does fine left out in a cool room. Don't wash it, eggplant doesn't like any extra moisture around its leaves. For longer storage, place loose, in the crisper.

Fava beans—place in an airtight container.

Fennel—if used within a couple days after it's bought fennel can be left out on the counter, upright in a cup or bowl of water (like celery). If wanting to keep longer than a few days, place in the fridge in a closed container with a little water.

Garlic—store in a cool, dark, place.

Green garlic—an airtight container in the fridge or left out for a day or two is fine, best before dried out.

Greens—remove any bands, twist ties, etc. Most greens must be kept in an airtight container with a damp cloth to keep them from drying out. Kale, collards, and chard even do well in a cup of water on the counter or fridge.

Green beans—they like humidity, but not wetness. A damp cloth draped over an open or loosely closed container.

Green tomatoes—store in a cool room away from the sun to keep them green and use quickly or they will begin to color.

Herbs—a closed container in the fridge will keep up to a week. Any longer might encourage mold.

Lettuce—keep damp in an airtight container in the fridge.

Leeks—leave in an open container in the crisper wrapped in a damp cloth or in a shallow cup of water on the counter (just so the very bottom of the stem has water).

Okra—doesn't like humidity. So a dry towel in an airtight container. Doesn't store that well, best eaten quickly after purchase.

Onion—store in a cool, dark, and dry place. Good air circulation is best, so don't stack them.

Parsnips—an open container in the crisper, or, like a carrot, wrapped in a damp cloth in the fridge.

Potatoes—(like garlic and onions) store in cool, dark and dry place, such as a box in a dark corner of the pantry; a paper bag also works well.

Radicchio—place in the fridge in an open container with a damp cloth on top.

Radishes—remove the greens (store separately) so they don't draw out excess moisture from the roots and place them in a open container in the fridge with a wet towel placed on top.

Rhubarb—wrap in a damp towel and place in an open container in the refrigerator.

Rutabagas—an ideal situation is a cool, dark, humid root cellar or a closed container in the crisper to keep their moisture in.

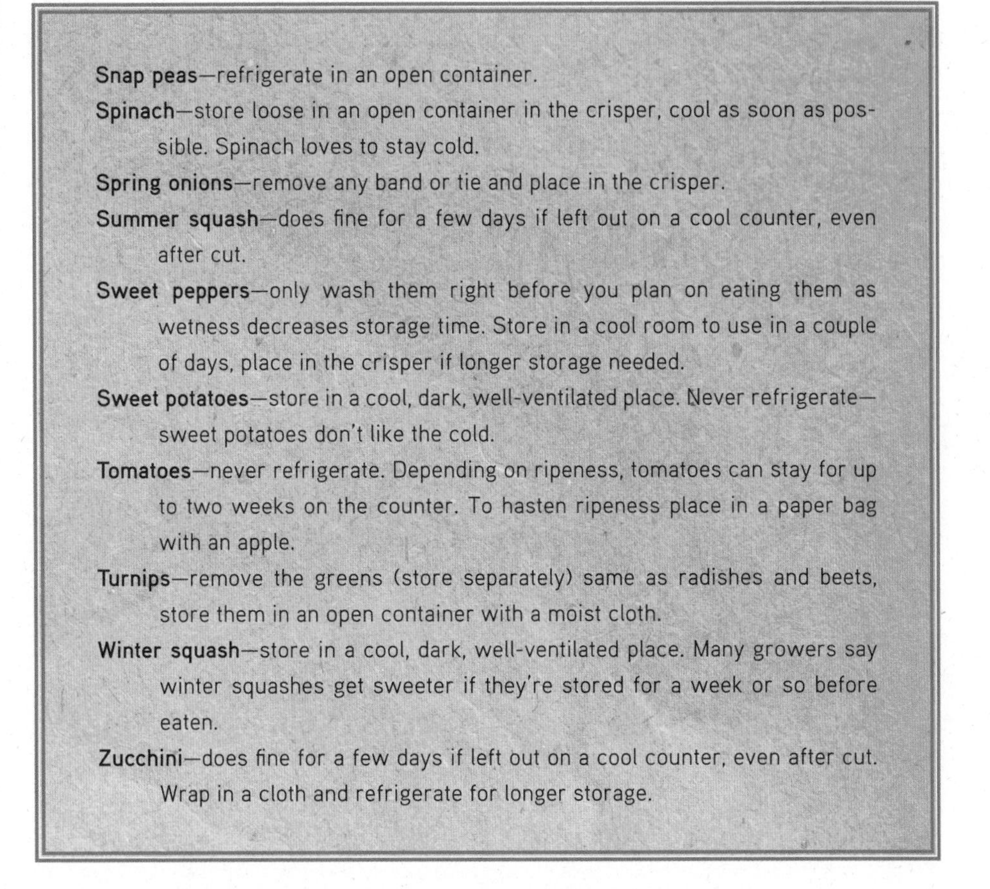

Snap peas—refrigerate in an open container.

Spinach—store loose in an open container in the crisper, cool as soon as possible. Spinach loves to stay cold.

Spring onions—remove any band or tie and place in the crisper.

Summer squash—does fine for a few days if left out on a cool counter, even after cut.

Sweet peppers—only wash them right before you plan on eating them as wetness decreases storage time. Store in a cool room to use in a couple of days, place in the crisper if longer storage needed.

Sweet potatoes—store in a cool, dark, well-ventilated place. Never refrigerate—sweet potatoes don't like the cold.

Tomatoes—never refrigerate. Depending on ripeness, tomatoes can stay for up to two weeks on the counter. To hasten ripeness place in a paper bag with an apple.

Turnips—remove the greens (store separately) same as radishes and beets, store them in an open container with a moist cloth.

Winter squash—store in a cool, dark, well-ventilated place. Many growers say winter squashes get sweeter if they're stored for a week or so before eaten.

Zucchini—does fine for a few days if left out on a cool counter, even after cut. Wrap in a cloth and refrigerate for longer storage.

If you find that the steps listed above don't keep your fruits and veggies fresh for long enough, see Chapter 9 for more ways to store foods without plastic.

Beautiful Bulk Bins

The biggest step in my plastic-free grocery shopping journey was learning to use and love the bulk bins. Most natural foods stores, as well as some conventional supermarkets, have a section of bins where you can buy loose unpackaged rice, grains, flours, soup mixes, beans, cereals, trail mixes, sugar, cocoa powder, chocolate chips, dried fruits, nuts, seeds, olives, herbs, spices, and teas. Some stores offer liquids in bulk, like cooking oils, vinegar, honey, maple syrup, and soy sauce. Nut grinders allow customers to grind their own peanut or almond butter into their own containers. Out here in the Bay Area, independent grocery stores like Rainbow Grocery and Berkeley Bowl provide an even wider selection of dried pastas and crunchy snack foods. Rainbow even offers bulk fresh pasta, pesto, miso, tofu, and many other products you might not expect to buy packaging-free. Here are a few suggestions for making friends with the bulk bins.

Bulk bins at Rainbow Grocery in San Francisco

Find Out What Resources Exist Where You Live. It might require a little traveling at first to research what stores in your area carry the widest selection of bulk products. You usually won't find much more than candy, nuts, and some dried fruits at conventional grocery stores like Safeway. Chains like Whole Foods Market and local natural foods stores and co-ops will be your best bet. Take notes when visiting these stores, so you'll remember what they offer and can plan future shopping trips to stock up on necessities.

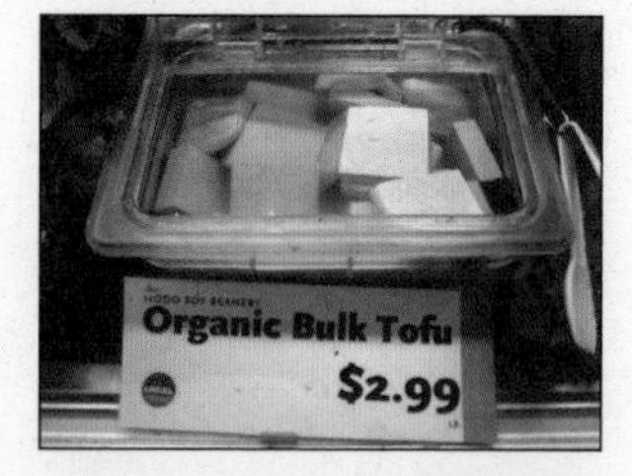

Bulk tofu from Whole Foods

Bring Your Own Bags and Containers. While proponents tout bulk bins as a way to generate less packaging waste, stores generally provide rolls of plastic bags for bringing home your purchases. Avoid the plastic bags. Just as you can use your own cloth produce bags

Bulk pasta from The Pasta Shop in Oakland

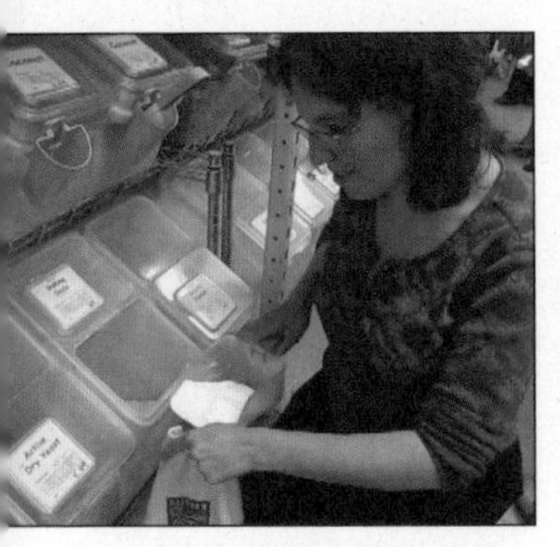

Bring your own bags to buy from bulk bins.

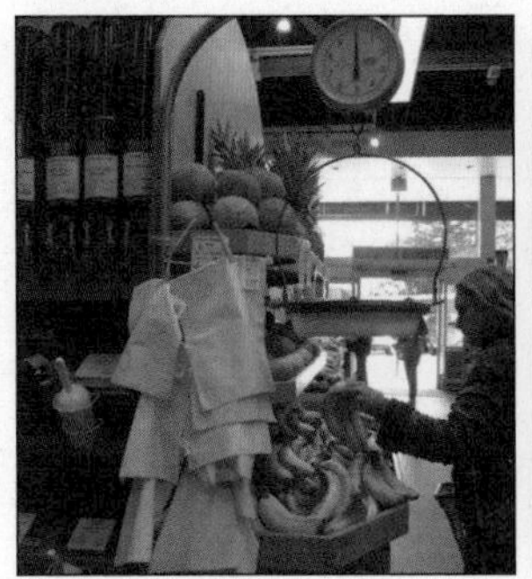

Cloth bags for produce or bulk foods

Bulk fresh pasta from Rainbow Grocery in San Francisco

for buying fruits and vegetables, many stores allow you to use your own reusable bags and containers for bulk foods. If the store doesn't sell reusable bulk/produce bags, you can buy them through Eco-Bags Products, ChicoBag, ReUseIt, or Etsy. Another company I love is Ambatalia, which offers gorgeous handmade, natural linen, multipurpose *bento* bags, designed to carry everything from produce to popcorn. Ambatalia's motto is "Hand-crafted textiles for a non-disposable life." I love that.

Alternatively, try making your own bulk bags out of repurposed t-shirts and pillow cases. Or skip the bag altogether and fill up repurposed glass mason jars, which means you don't have to pour the contents of the bag into a storage container when you get home. Pretty much any container will do, as long as you follow the next step.

Weigh Your Containers/Bags. Bulk foods are sold by weight. You don't want the weight of your bags and containers to be included in the price of the food, so make sure you weigh your containers before you fill them. Different stores have different methods for dealing with the "tare" weight. At Whole Foods, customers first bring their containers to the Customer Service Desk, where they are weighed by a staff member, who writes the weight on a sticker. Then, at checkout, the clerk deducts the weight of the container. Rainbow Grocery, on the other hand, uses the honor system. Customers weigh their own containers on scales provided in the back of the store and mark the weight on the container, which is deducted at checkout. If you haven't used the bulk bins before, ask store staff the proper way to handle tare weights as soon as you enter the premises. And by the way, most reusable bulk/produce bags offered for sale will indicate the tare weight on the label, so you don't have to worry about weighing them.

> "Since I use my own storage containers to buy bulk foods, I always know what I need! No need to make a shopping list—my list is all of my containers!" —Danielle Richardet, www.itstartswithme-danielle.blogspot.com, Wilmington, North Carolina

A Funnel Can Help. I like to bring a funnel with me if I know I'm going to be filling up glass jars with narrow necks. Invariably, another customer will ask to borrow it, and we'll end up having a conversation about buying in bulk and reducing waste.

Reuse Your Plastic Bags. Some stores, unfortunately, do not allow customers to use their own bags and containers. And some stores are not set up to deduct tare weights. In that case, you can bring back plastic bags to reuse again and again. Once, while shopping at Berkeley Bowl, I saw a woman who had devised an ingenious system for reusing her plastic bags. Each bag was labeled with the name of a product and filed alphabetically in a folder. As she used up products, she'd file her bags in a shopping folder that she brought to the store with her. The system was her shopping list and also allowed her to avoid constantly washing out bags that had contained dry foods. Why rinse out her white flour bag, for example, when she's just going to refill it with white flour? If you do choose to reuse plastic bags, be sure and empty the contents into a nonplastic container as soon as you get

A funnel is helpful when filling your own jar from the bulk bins.

ZERO-WASTE HERO:
Tracey TieF

A few years ago, my friend Tracey called from Toronto to tell me she was coming to the Bay Area on her way to the annual Burning Man event out in Nevada's Black Rock Desert and wanted me to take her grocery shopping to stock up for the trip. No, she didn't want to go to just any grocery store; she wanted to buy all her supplies from the produce aisles and bulk bins. I was excited to show her the resources we have here in the Bay Area and not at all surprised that Tracey—who in her holistic health practice (Anarres Natural Health in Toronto) sells handmade personal care products in glass bottles and gives her customers discounts for bringing back their bottles to refill—would want to shop this way. What I didn't expect was the extra red suitcase she'd toted all the way across the continent just for the shopping trip. (She told me she'd actually packed it full of clothes inside a bigger suitcase, which was also packed, lest I think she paid to check an empty suitcase on the plane.) Tracey was prepared. For a year she'd been saving up the unavoidable plastic baggies in which her raw ingredients sometimes come packaged. A few of them still smelled like peppermint essential oil, but Tracey was committed to reusing them anyway.

I took her to Berkeley Bowl, where she filled her baggies and a few stainless steel containers with dry soup mixes (black bean, split pea, curry lentil, corn chowder), hummus mix, falafel mix, dried tabouleh, dried fruit, nuts, crackers, soy milk powder, and whatever else could easily be prepared while camping out in the desert. And she bought extra food to share, since that is what you do at Burning Man. Then, after checking out, she filled up that red suitcase and wheeled it out of the store, but not before I took a photo to show people how bulk food shopping can be done. After several years of buying in bulk and avoiding new plastic, I thought I was an expert on plastic-free living, but that day, Tracey became my hero and role model. How many of us are not only willing to maintain our eco-practices while on vacation but come prepared to follow through?

home. While reusing plastic bags is one way to avoid consuming new plastic, we would be better off not having to use plastic bags in the first place. While you're at the store, why not talk to the manager about the policy and let him/her know why going plastic-free is best for our health and that of the planet?

Take Some Bags with You Whenever You Leave the House. Successfully relying on bulk bins to reduce plastic waste does require planning. You can't just pop into the nearest store for sugar or cereal if you don't have bags or containers with you to carry them home. I keep a few cloth bags and even some reused paper bags in my purse at all times, just in case.

Plan Your Shopping Trips. If you live more than a few miles away from a bulk foods store, plan big shopping trips so you can stock up all at once and reduce travel miles. My friend Lisa Sharp, the Oklahoma recycling hero from chapter 4, buys most of her food from a natural grocery store many miles from her house. But she and her husband plan ahead, so they only have to go there once a month.

Learn to Cook with Basics from the Bulk Bins. Dried legumes are a staple in our house, especially since I discovered how quickly they can be prepared in a pressure cooker. Instead of opening a BPA-lined can, we've learned to make homemade hummus from dried chick peas and regularly eat lentils, split peas, black beans, and pinto beans. (Yes, I've come a long way since my frozen dinner days!) Whole Foods Market provides a handy booklet, available near the bulk bins, that explains how to prepare most of the products sold in bulk. (You can also find the information on the Whole Foods website.) But my favorite resource is *The Everything Beans Book*, an e-book from my blogger friend Katie, which you can download via her website www.kitchenstewardship.com. Preparing beans is not hard at all; it just takes time because beans should be soaked before cooking.

Are Bulk Bins Safe? People often ask me whether purchasing food from bulk bins is sanitary. What if little children stick their hands in bins? For that matter, what if grownups—who should know better—do it? What about people with allergies who must avoid cross-contamination? And how are the foods packaged ***before*** they go into the bulk bins? To learn the answers to these questions, I met up with Denise Jardine, healthy eating coordinator for Whole Foods Market, Northern California region. Here's what she told me:

- **Bins, Tongs, and Scoops Are Sanitized Regularly.** Each Whole Foods store has its own commercial dishwasher and heavy duty sanitization system that is audited monthly by a third party. The bins are on a cleaning rotation. Each night, three to eight bins are selected for sanitizing, so each bin ends up being thoroughly cleaned about once a month. Tongs and scoops are sanitized nightly. If you shop at a store other than Whole Foods, why not ask the manager how the bulk section is kept clean?
- **Signs Remind Customers to Keep Their Hands Out.** Signs on the Whole Foods bulk bins read: NO HANDS! PLEASE USE SCOOP OR TONGS IN BINS. Of course, signs don't guarantee that wee ones won't stick their hands in. Parents, please pay attention to your offspring! If you're really concerned about germs, you can limit your bulk purchases to foods from overhead bins that are released by gravity rather than being scooped. You could also choose foods that must be cooked, such as beans, rice, grains, pasta, or flour. But personally, I buy nuts, dried fruit, candy, pretzels, and other snacks from bulk bins and feel that any germs present are only helping me boost my immune system.
- **Designated Scoops Help Prevent Cross-Contamination.** Scoops and tongs are attached by a cord to the bin in which they are meant to be used. Still, no cross-contamination cannot absolutely be guaranteed. Denise, who knows a thing or two about food sensitivity since she wrote the book ***The Dairy Free & Gluten Free Kitchen*** (Ten Speed Press), recommends that people with life-threatening allergies or celiac disease stick to prepackaged foods, but those with sensitivities that are not so extreme should be safe buying anything that can be rinsed. Flour would not be a good candidate, but rice or beans might be.
- **Some Bulk Foods Are Shipped in Plastic.** Foods like rice, whole grains, and flour come packaged in great big paper bags. But sugar is packaged in plastic bags. Nuts, dried fruits, and other products generally come in big card-

board boxes with plastic liners. So while buying in bulk doesn't completely eliminate plastic food packaging, it does reduce it significantly. Like I said in the beginning of this chapter, buying large sizes reduces the packaging to product ratio.

One Company is Going Further. While many natural grocery stores provide a selection of foods in bulk, most of their shelf space is dedicated to packaged foods, and most of that packaging is plastic. Recognizing the unsustainability of all that packaging waste, Austin, Texas-based grocery store in.gredients has developed a new way to do business.

When I wrote about in.gredients in the first edition of this book, the store had not yet opened. The founders were planning to create the first nearly packaging free grocery store in the United States and hoped to eliminate packaging throughout the supply chain, including the packaging most customers don't ever see. Because they would source their products locally, they could encourage their vendors to deliver products in reusable, returnable containers, which would be swapped out with each new delivery. So I was excited to meet in.gredients founder Christian Lane at a Think Beyond Plastics event in 2013 and even more excited to actually visit the store in October of that year. in.gredients turned out to be better than I expected.

First of all, they have developed the most high tech tare system I've ever seen: a scale attached to a computer, monitor, and label printer is located right inside the front door, so there's no mistaking that in.gredients wants you to use your own containers. You weigh your container (I happened to have a Lunchbots container with me), press the button on the screen, and a sticker prints out with the tare weight and a bar code. Apply the sticker to your container and then fill it with one of the many bulk items in the store. Write the item's code on the container with a provided grease pencil (which can be washed off easily.) When you check out, the cashier enters the code,

Quick Tip:

For more information about using bulk bins or for help encouraging your local store to carry more bulk items, visit the Bulk Is Green Council at www.bulkisgreen.org.

weighs your container, scans the bar code, and the system automatically deducts the tare weight. The bar code can also store information about how many times you've used that container. (The sticker is made from a high tech material that can withstand infinite dishwashing cycles without coming off.)

What can you put in your containers? Oh, so many things . . . herbs, spices, teas, nuts, grains, beans, pasta, oils and other "wet" foods, personal care products, soaps and other cleaners . . . even natural insect repellent. Yes, I discovered, Austin has mosquitoes. in.gredients also sells tasty beverages on tap. Beer, wine, kombucha, soda . . . just fill up a returnable jug or bottle. (Unlike those in most other states, grocery stores in Texas are allowed to sell beer and wine in customers' own containers. in.gredients also offers a delicious selection of prepared foods, which you can take to go, or stay and enjoy on a durable plate with a real glass for your drink.

The only foods not sold unpackaged are meat and dairy because the store does not have a butcher counter or cheese counter but still wants to support local organic meat and dairy producers. Still, I noticed that the cheeses were not individually cut and wrapped. To save packaging, cheese is cut to order and then wrapped in paper (which is required by local regulations).

But like I said, in.gredients's original goal was to eliminate packaging from the entire supply chain. And indeed, in the back room, I saw many empty containers waiting to be picked up and refilled. During his presentation at the Think Beyond Plastics conference, Christian Lane said, "It's important to educate customers." Every day, in.gredients does just that.

If I were a vegan (a lofty goal I've aspired to for years but have yet to reach), I could feed myself quite well just with foods from the produce section and bulk bins. But I've found ways to buy all kinds of other foods plastic-free. Keep in mind as you read the rest of this chapter that it's taken me years to get to this point. Don't try to do it all at once! While some of my solutions may be easy changes to make, some of them may seem extreme. It's up to you which ones you decide to tackle. I provide these ideas as a resource for you to see what's possible, as well as what roadblocks still exist and how we need to change the system to make plastic-free grocery shopping easier for all of us.

Bread & Baked Goods

Bake It Yourself. If you like to make bread, this one is a no-brainer. Just buy flour from the bulk bin and bake it yourself. In my area, I can even buy the yeast in bulk, but you can also find yeast in glass jars as opposed to individual packets lined with plastic. With a bread machine, the task is even easier. Unfortunately, I don't have a bread machine, and anticipating how little I'd probably end up using it, I'm not inclined to buy one—even secondhand. My attempts at making whole wheat pita bread a few years ago were, well . . . imagine little brown Frisbees that would give you a concussion if they flew into your head. I just don't eat pita anymore. But many people are solving the plastic bread bag problem by finding ways to make it themselves. My blogger friend Danielle has posted recipes for homemade tortillas and burger buns (www.itstartswithme-danielle.blogspot.com). And all over the Internet, homemade bread recipes abound. Google is your friend. Go for it!

> "Try to align your hobbies with your desire to live plastic-free. Baking, for example, will help you cut way back on cookie packaging, bread bags, etc. The same can be said for woodworking, sewing, canning, gardening—they're all hobbies that can help you cut back on plastic while learning new skills and having fun." —Ellen Simpson, Salem, Massachusetts

Buy It in Paper or Packaging-Free. Fortunately, in my area, there are several artisan bakeries that sell loaves of unsliced bread in paper bags instead of plastic. I keep the paper bags and reuse them. But better still are the bakeries in my neighborhood with shelves full of "naked" bread that are happy to put loaves directly into my own cloth bulk bag. Are there bakeries in your area that sell bread in paper or without any packaging at all? Have you checked? A representative from one U.S. bakery chain, Panera, told me they will place unwrapped bread directly into a customer's cloth bag as long as it is clean.

Don't Slice It Until You're Ready to Eat It. Bread stays fresher when it's kept intact. I buy it unsliced and slice as I go.

Buy bread
without plastic.

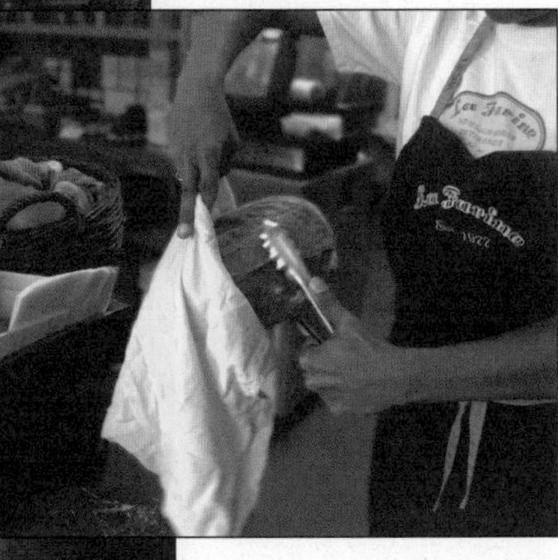

Homemade "wheat thins"

Store Bread in Cloth Bags in an Airtight Container. I have a stainless steel bread box on my kitchen counter that keeps bread fresh for several days. Bread stays fresh even longer in the refrigerator in a cloth bag inside Life Without Plastic's 1-gallon rectangular airtight container. I know there are people who say you shouldn't keep bread in the refrigerator, but this is really the very best method I have found for keeping it both soft and mold-free. You can make your own breadbox from an old popcorn or biscuit tin. Thrift stores are full of them. But be aware that those tins will rust if exposed to moisture. Stainless steel is expensive, but it lasts indefinitely.

Bag Your Own Bagels and Donuts. Many conventional grocery stores offer unpackaged bagels and donuts in a glass case. Bring your own bags and containers to take them home.

Plastic bread bag clips litter the ground around Oakland's Lake Merritt.

Bring Your Own Container for Cakes, Cookies, and Other Sweet Treats. Skip prepackaged cookies or cakes in plastic containers and opt for fresh-baked goodies from the bakery counter. Ask the staff to use your container.

Make Your Own Snacks. You don't have to give up crackers, energy bars, and other snacks that come packaged in plastic if you learn to make them yourself. Once again, Google is your friend. But I didn't seriously consider making my own crackers until my friend Katie came out with her *Healthy Snacks to Go* e-book, which even contains a recipe for homemade "Wheat Thins." I made them. They turned out great—better than the store-bought version. You can download the e-book from Katie's site, www.kitchenstewardship.com.

Be Careful of Plastic Bread Bag Clips. When considering which bread to buy, keep in mind that the plastic packaging includes not just the plastic bag but often a little plastic clip that keeps the bag closed. I find these little clips scattered over the ground near Lake Merritt in Oakland, where visitors have dropped them after feeding the ducks and birds loaves of bread. They are not only a hazard to animal life, but also elderly humans. A strange article published in the *Canadian Medical Association Journal* in 2000 reported doctors finding plastic bread bag clips inside the gastrointestinal tracts of people over sixty years of age who wore full or partial dentures. The article did not specify how people accidentally swallow the clips—I imagine it's from holding a clip in the mouth temporarily while taking out a slice of bread—but it did show some pretty graphic harm as a result.

Twist ties, on the other hand, may be coated with PVC. The point is to be mindful of all the plastic associated with food packaging, especially small items that are often overlooked.

Quick Tip:

Plastic-Free Amish Friendship Bread It's the chain letter of baking: friends pass on the yeast "starter," usually in a plastic Ziploc bag, which they "feed" with additional flour, sugar, and milk, quadrupling the amount each time. Then, they keep one batch and pass the rest on in more plastic bags. But we can put an end to the plastic bag monster and still participate in the fun. Use any non-metal container and a wooden spoon. Distribute the new starter in glass jars instead of plastic baggies. Remember—the Amish wouldn't have used plastic, and you don't have to either.

Dairy Products and Soy Milk

Choose Milk in Returnable Glass Bottles. The truth is that all milk is packaged with some kind of plastic, whether it's an HDPE plastic bottle or jug, a paperboard carton coated with polyethylene, a milk box with a plastic liner, or even a glass bottle with a plastic cap. The glass bottle is the best option, plastic cap notwithstanding, because the milk isn't encased in plastic and because usually the bottles can be returned to the store for refilling rather than recycling. Customers pay a deposit to the store when they buy the milk and get the money back when they return the bottle. You'll find a list of dairies using glass milk bottles at www.mindfully.org/Plastic/Dairies-Glass-Bottles-Milk.htm. The page has not been updated since 2009, so if you don't find glass-bottled milk in your area on the list, try Google.

If you don't have access to milk in glass bottles where you live, contact your local recycling facility and find out which packaging is most easily recycled in your area. Plastic-coated cardboard cartons may be harder to recycle than HDPE plastic milk jugs and are problematic for many composting facilities. You may not be able to avoid plastic entirely, but you can at least make sure that what you do choose doesn't end up in the landfill.

Buy milk in glass bottles.

Buy Cheese in Bulk. To buy cheese with less plastic, take a reusable container or wrap to the cheese or deli counter and ask the clerk to use it instead of wrapping your cheese in plastic. I call this procedure *less*-plastic cheese buying because the original block or wheel of cheese will be wrapped in plastic as soon as it is cut. If you are up for the challenge of going completely plastic-free, find a cheesery that sells whole wheels of cheese without plastic packaging and buy the whole wheel. The upfront cost can be high, of course, so consider splitting it with friends. Out here, I buy wheels of Bellwether Farms Carmody (which to me tastes sort of like Monterey Jack) that comes without any plastic wrap.

Store Cheese without Plastic. Once you have your plastic-free cheese, how do you store it so it doesn't grow mold or dry out? A small amount can be wrapped in a cloth and stored in an airtight container in the refrigerator. For a whole wheel, there are several options.

- **Option 1:** Store entire wheel wrapped in cloth in an airtight container in the refrigerator. Rub the cut face of the cheese with olive, canola, or other vegetable oil. If mold starts to form, it will consume the oil instead of the cheese. Simply wipe it off or rinse in tepid water. Dry and rub with fresh oil and store again.
- **Option 2:** Shred cheese into glass jars or other airtight containers and store in the freezer. This is my preferred option these days, once I discovered the grater attachment on my food processor. Now I don't have to worry about cheese going bad or drying out, and the shredded cheese thaws very quickly. Apologies to cheese connoisseurs who might be appalled at the idea of freezing cheese.

> "Making your own almond milk is really easy. Just soak 1 cup of raw almonds in 4 cups of water overnight, blend it in the blender, and strain it through a dish towel. You can throw in a couple of dates to sweeten it. I put my almond milk in smoothies and on homemade granola." —Mary Katherine Glen, www.brokebusyeco.wordpress.com, Mountain View, California

Make Your Own Yogurt or Buy It Plastic-Free. It's easy to make yogurt even if you don't have a yogurt maker. See page 205 for the method I have used successfully with a regular thermos. The key is to find a way to keep the yogurt at the right temperature while the cultures are doing their work. If you don't want to make your own, you may have other plastic-free options. In the Bay Area, the yogurt company St. Benoit sells its yogurt in glass jars. Do any yogurt companies near you package their yogurt without plastic?

If plastic is your only option, buy the biggest size container you can, and skip the single-serving sizes. Also, contact yogurt companies and ask whether they fill the containers while the yogurt is still hot. Most large companies do, and the hotter the yogurt, the more likely the plastic is to leach chemicals. The Bay Area's Straus Family Creamery *vat sets* its yogurt in a big stainless steel tank and allows it to cool down before filling the plastic containers.

Make Your Own Soy or Nut Milk. Whether from the shelf or the refrigerator, soy milk comes in plastic-coated cartons, usually with a plastic spout, or worse, in aseptic cartons with their multiple layers of plastic, paper, and aluminum. If you drink soy, nut, or seed milks regularly, consider either investing in a soy milk maker or learning to make it with your blender. Google "How to make homemade soy milk." (If you do go the blender route, you'll need to strain your nut milk. Check out the organic cotton and hemp nut milk bags from Eco Peaceful. Not only is the fabric made from natural fibers, but the seams are sewn with organic cotton or linen thread without the use of glues to secure them.) I bought the SoyaPower machine from Sanlinx (www.soymilkmaker.com) in the early days of my plastic-free project because I figured that even if the machine contains plastic, it will save a lot of packaging waste in the long run. Sadly, I have to confess that not long thereafter, I got tired of cleaning out the soy milk maker after each use and it became just another gadget gathering dust on my kitchen counter, not doing me or the environment any good. Finally, I sold it locally via Craigslist. Hopefully the buyer will get more use out of it than I did. So you see why I'm reluctant to buy a bread machine. I do learn from my mistakes.

Michael brings a stainless steel bucket to the butcher counter to buy meat for our kitties.

Meat & Deli

While we don't eat much meat in our home, our kitties are ravenous carnivores. We buy our meat from the butcher counter at Whole Foods Market or from the independent butcher shop up the street using our own stainless steel pot with an airtight lid. The butcher can weigh the container and deduct the tare weight from the final price, just as is done with bulk foods. I love the surprised looks we get from other customers and the frequent comment, "Wow. That's cool. I didn't know you could do it that way." You don't know until you ask! We buy salads and cooked foods from the deli/prepared foods counter the same way.

Homemade Thermos Yogurt

Thanks to Melanie Rimmer, whose blog Bean Sprouts gave me the inspiration years ago.

1. Fill a thermos with whatever kind of milk you want to use. (This step is just to measure out the milk.)
2. Pour the milk into a pan or microwave bowl and bring to a boil.
3. Remove from heat or microwave and stick a thermometer into the milk. Allow the milk to cool to 122°F (50°C).
4. Pour milk back into thermos and add a tablespoon (15 mL) of yogurt from a previous batch. If this is your first batch, you'll have to use a tablespoon of store-bought yogurt, but in the future, you can keep your batches going using your own yogurt. Make sure if you use store-bought yogurt that it still contains active yogurt cultures.
5. Cover thermos and let sit for 8–14 hours.
6. For less watery yogurt, strain it through a wire sieve over a bowl.

"One day I purchased a pork shoulder from the meat deli at our co-op, and I was appalled when I got home, unrolled it, and saw how much paper and plastic waste there was. My next trip to the co-op, I was equipped with my snap glass and Pyrex storage. It takes some planning, but if you know what you are going to be purchasing, it's not a problem." —Amber Husten, www.thetastyalternative.com, Davis, California

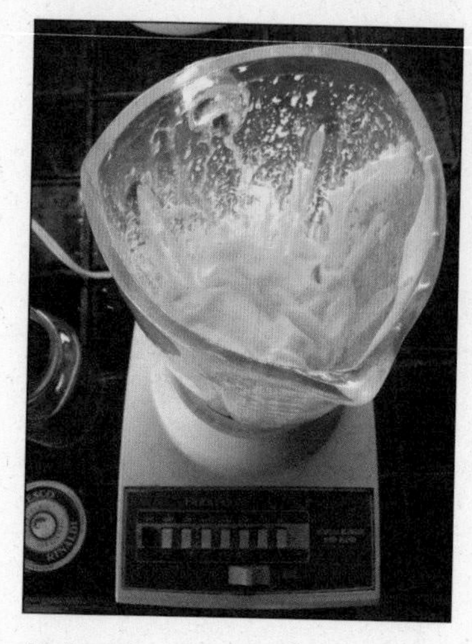

Making homemade mayo

Condiments, Spreads, Sauces, & Other Processed Foods

There's Almost No Food You Can't Make Yourself. Many processed sauces, spreads, and condiments are available in glass jars instead of plastic. But before opting for energy-intensive glass, I look for ways to make my own and skip the packaging altogether. By searching the Internet, it's possible to find a recipe for almost any processed food you can think of: fruit leather, ketchup, pasta sauce, tomato paste, hot sauce, salsa, guacamole, bean dip, hummus, Indian simmer sauces, you name it. They differ in complexity and length of time required. I certainly don't make everything from scratch, but I do find it fun to experiment and try to make homemade versions of common processed foods at least once.

Homemade Chocolate Syrup

Plastic-free living is not about deprivation. My absolute favorite discovery is plastic-free chocolate syrup, which is a staple in our home. We put it in milk. We pour it on cake. The one thing we don't do anymore is squeeze it from a plastic bottle. It's really easy to make from scratch and tastes a lot better than store-bought.

Ingredients:

- 1 cup (120 g) unsweetened cocoa powder (bulk bin)
- 1 cup (180 g) brown sugar (bulk bin)
- 1 cup (200 g) raw sugar (bulk bin)

- ¼ teaspoon (1 g/1.23 mL) salt (bulk bin or cardboard container)
- 1 cup (240 mL) cold water (tap)
- 1 tablespoon (15 mL) vanilla (glass bottle)

Combine cocoa and sugar in a saucepan and blend until all lumps of cocoa are gone. Add water and salt and mix well. Cook over medium heat, bringing to a boil, stirring constantly. You want to make sure there are no lumps! Continue stirring on the stove for just a couple more minutes, being careful not to let the sauce burn on the bottom of the pan. The sauce should still be fairly runny. Remove from heat and let cool. The sauce will thicken up as it cools. Add vanilla.

We store our chocolate syrup in a small ceramic pitcher in the refrigerator.

Note: This is syrup, not fudge sauce, and it will not be as thick as fudge sauce. It's great for chocolate milk, hot cocoa, and topping ice cream and cake. Feel free to experiment with different kinds and amounts of sugar or flavoring extracts. Let your taste buds be your guide.

Homemade Mayonnaise

My late friend Norma Draper shared this recipe with me. She swore that ever since she got her very first Waring blender—which is a long time because she lived into her eighties—she never bought another jar of store-bought mayonnaise. Why would she pay for it when homemade is so easy and inexpensive?

Ingredients:

- 1 whole egg (cardboard carton)
- 2 tablespoons (30 mL) vinegar or lemon juice (glass bottle or fresh-squeezed)
- ½ teaspoon (2 g/2.46 mL) dry mustard (bulk bin or glass jar)
- ½ teaspoon (1 g/1.23 mL) salt (bulk bin or cardboard container)
- 1 cup (240 mL) salad oil (bulk or glass bottle)

Place egg, vinegar or lemon juice, seasonings, and ¼ cup (60 mL) of the oil in the blender in the order indicated. Put on cover. Run blender until contents are thoroughly blended, about 5 seconds. Remove cover. Add remaining oil very gradually, running blender as you add oil. ***Do not add oil all at once!*** Run blender for a few seconds more after last oil is added. Store in a glass jar in the refrigerator. Yield: About 1-¼ cups (300 mL).

According to several websites, you can also make mayonnaise with an immersion blender or even a wire whisk, which takes much longer but saves electricity and avoids the blender's plastic parts.

Mark P's Homemade Ketchup

Another friend, Mark Peters, makes most of his own condiments and invited me over one foggy December afternoon to hang out and learn how to make homemade ketchup. We spent several hours chatting and relaxing because while homemade ketchup is easy to prepare, it requires several hours of cooking time.

Note: While some homemade ketchup recipes online call for processed tomato paste, which often comes in BPA-lined cans, this recipe uses only fresh ingredients.

Ingredients:

- 4 pounds (1.8 kg) tomatoes (farmers market, produce aisle, garden)
- 1 large onion, chopped (farmers market, produce aisle, garden)
- 1 cup (140 mL) your choice of vinegar—Mark uses plain white. (bulk or glass bottle)
- 1 teaspoon (5 g/5 mL) salt (bulk bin or cardboard container)
- 1 teaspoon (5 g/5 mL) ground cloves (bulk bin or glass jar)
- 1 teaspoon (5 g/5 mL) ground allspice (bulk bin or glass jar)

Drop tomatoes into a pot of boiling water for about a minute until their skins split. Once skins have split, the peel will basically fall off. Peel and chop tomatoes. Combine with chopped onions in a large saucepan and simmer for about 10 minutes. Transfer tomato/onion mixture in small batches to a blender with a glass pitcher (I don't recommend putting hot foods into a plastic blender pitcher!) filling it only about half full each time. Puree each batch and pour into a bowl.

When finished pureeing, pour the entire batch back into the saucepan, making sure there are no more big chunks. Add vinegar, salt, cloves, and allspice, and stir. Let the ketchup simmer slowly, uncovered, for several hours, stirring occasionally, until it is reduced about 50 percent or to the desired thickness. Mark's ketchup ends up a brownish red color, not the artificially enhanced red of many commercial ketchups. But believe me, it tastes fantastic. You can try adding a little lemon juice or sugar to preserve the color.

Transfer ketchup to jars and let cool before refrigerating or freezing. It will keep for about four months in the refrigerator and indefinitely in the freezer. **Important**: If you plan to freeze the ketchup, do not fill the jar all the way. Leave space at the top for expansion. Glass jars are fine in the freezer as long as they are not overfilled.

Homemade Spicy German Mustard

This recipe is super easy but takes much longer than mayonnaise and requires more ingredients. It lasts a long time in the refrigerator.

Ingredients:

- ¼ cup (45 g) yellow mustard seeds (bulk bin or glass jar)
- 2 tablespoons (22 g) black or brown mustard seeds (bulk bin or glass jar)
- ¼ cup (30 g) dry mustard powder (bulk bin or glass jar)
- ½ cup (120 mL) water (tap)

- 1½ cup (350 mL) cider vinegar (glass bottle or in bulk if you can find it)
- 1 small onion, chopped (farmers market, produce aisle, garden)
- 2 tablespoons (23 g) firmly packed brown sugar (bulk bin)
- 1 teaspoon (5 g/5 mL) salt (bulk bin cardboard container)
- 2 garlic cloves, minced or pressed (farmers market, produce aisle, garden)
- ½ teaspoon (2 g/2.46 mL) ground cinnamon (bulk bin or glass jar)
- ¼ teaspoon (1 g/1.23 mL) ground allspice (bulk bin or glass jar)
- ¼ teaspoon (1 g/1.23 mL) dried tarragon leaves (bulk bin or glass jar)
- ⅛ teaspoon (0.5 g/0.5 mL) turmeric (bulk bin or glass jar)

Combine the mustard seeds and powder in a small bowl. Combine the remaining ingredients in a stainless steel, or other nonreactive, saucepan. Simmer the mixture uncovered on medium heat until reduced by half. Combine with mustard mixture in bowl. Cover bowl (I place a saucer on top of the bowl) and let stand at room temperature for 24 hours. (Add additional vinegar if necessary to keep the seeds covered.) Process the mixture in a blender or food processor until pureed to the texture you like. Scrape mustard into clean, dry jars. Cover tightly and age at least 3 days in the refrigerator before using. The mixture will continue to thicken. If it gets too thick after a few days, stir in additional vinegar. Makes about 1½ to 2 cups (350-475 mL).

Note: Initially, the mustard will be hot enough to light your face on fire. The longer it ages in the refrigerator, the milder it will become.

Homemade Pesto

Pesto is one of the easiest sauces to make because it's very hard to mess up. Here are the basics, without exact ingredient amounts because I pretty much just taste as I go, adding this and that until it tastes good to me.

Ingredients:

- A bunch of leaves (farmers market, produce aisle, garden) These are usually basil, but you can use arugula or other aromatic leaf.
- A bunch of nuts (bulk bin) Traditionally, pesto is made with pine nuts. But I prefer almonds. You can also use walnuts, or anything else that seems like it would taste good.
- Olive oil (bulk or glass bottle)
- A little lemon juice (fresh-squeezed or glass bottle)
- A couple of garlic cloves (farmers market, produce aisle, garden)
- Options: parmesan cheese (if you can find it without plastic), hot chili flakes, salt, whatever your heart desires.

Process in food processor, testing as you go. If it's too thick, add more olive oil or lemon juice. If it's too watery, add more nuts. With practice, you learn what ingredients to add to make it perfect. I've never found a store-bought pesto that tastes better than what I can make at home. The hardest part is cleaning the food processor when I'm done.

Choose Foods in Glass. When you don't have the time or inclination to make foods from scratch and you can't find them in bulk, choose foods in safe, nonleaching glass instead of plastic or BPA-lined metal cans. It's true that glass is heavy and requires more energy to ship, but for food, it's safer than plastic. Foods I have found in glass include peanut and other nut butters (Once Again Nut Butter, Artisana); tomato paste (bionaturae); crushed tomatoes (Eden Organic); oils (Spectrum Organics, Artisana coconut oil); concentrated soup base (Better Than Bouillon); lemon and lime juice (Lakewood, Santa Cruz Organic); as well as pickles, olives, capers, horseradish, pasta sauces, vinegar, soy sauce, salsa, apple sauce, juices, honey, maple syrup, and many other products.

If you do choose glass, keep your jars and reuse them over and over to store leftovers. You can even use them in the freezer, which I'll discuss more in chapter 9. These days, I can't imagine putting something as useful as a glass jar in the recycle bin.

I try to find jars and bottles with metal lids, which are more recyclable than plastic. But bear in mind the next tip . . .

Metal Cans and Metal Lids Often Contain BPA. As I mentioned in chapter 1, the majority of metal food and beverage cans are lined with BPA, and state legislation banning BPA generally only focuses on the BPA in plastic bottles and sippy cups, rather than metal cans. What's worse, foods are processed at high heat, which makes BPA leach even more readily. I try to avoid canned foods these days, opting for glass when I can't find foods in bulk or am not willing to make them myself. But another sad truth is that the metal jar lids and screw caps on glass containers are often lined with BPA as well. How much BPA can leach into food from the lid is unclear. It's certainly a lot less than the amount of BPA that can leach into foods from the lining of metal cans. Personally, I don't go out of my way to avoid metal lids, since they are more recyclable than plastic lids and since we don't know what chemical additives are in the plastic lids anyway. What's needed is legislation banning BPA from all food containers so we don't have to worry about it.

Avoid Plastic in Tea Bags

Until recently, I assumed—like most people—that tea bags are compostable. I mean, they're made from paper, right? Not necessarily. Some special silken tea bags are made from nylon or, more recently, corn-based PLA plastic. But even many paper tea bags contain plastic fibers mixed in with the paper fibers to help seal the tea bags. Celestial Seasonings is one brand that confirmed for me that their paper tea pillows are infused with polypropylene fibers.[100] Other tea companies were not so forthcoming. To avoid the problem of plastic in tea bags—and reduce waste in general—look for loose teas in bulk. I buy my bulk tea from Rainbow Grocery in my own glass jar.

Wine and Beer

In some parts of the world—and some US states—you can bring your own container (i.e. growler) to fill up with wine and beer on tap. And of course, when patronizing a restaurant or bar, it's easy to order beer, and increasingly wine, on tap and avoid all of the waste associated with beverage packaging. (Check TryWineOnTap.com for a list of establishments pouring wine from stainless steel kegs rather than individual bottles or boxes. You can also add a location to the map.)

Unfortunately, California only allows wineries to fill growlers for off premise consumption, and since I don't live in wine country, I opt for glass bottles when I want to have wine at home. But wine-related plastic is not limited to the bottle itself but also the stopper and capsule—that wrapper around the neck of the bottle. I look for wines with natural cork stoppers, rather than plastic. It can be hard to tell just from looking at the bottle, but now there's a new mobile site at Corkwatch.org that can help. Corkwatch contains a huge database of wines and what kind of stopper they have: plastic, screw cap, or natural cork. You can look up existing wines or contribute to the site by adding smaller wineries and vintages that might not be listed. I also look for wines without any capsule around the neck at all. While some capsules are made from tin, more and more of them are made from PVC or plastic/aluminum laminates. And you can't always tell what the material is before you bring it home. Out here, there seems to be a movement among wineries to omit that wrapper entirely. I don't know if they leave it off for environmental reasons or simply to save money, but I appreciate any effort at less packaging.

A note about cork forests: Plastic wine corks are not only a problem because of all the usual issues with plastic, but they compete with natural Mediterranean cork oak forests, which not only provide humans with stoppers for their various libations, but also provide unique habitat for some of the world's unique and endangered animals, such as the Iberian Imperial Eagle and Iberian Lynx. Cork is harvested from the bark of trees, and when done properly, the process does not harm the trees but allows them to thrive and provide jobs for the people of the region. The World Wildlife Fund has a campaign to save Mediterranean cork forests. (You can read more about it on their website.)

Quick Tip:

If you're planning to make pumpkin pie for Thanksgiving, consider buying a whole pumpkin in October and storing it in a cool, dry space until November. Many mainstream grocery stores stop selling real pumpkins after Halloween and offer only canned pumpkin for your Thanksgiving feast.

Chewing Gum: It's Plastic

That's right. Chewing gum is made from plastic. If the label simply lists "gum base" as an ingredient, it may contain polyethylene, polyvinyl acetate, petroleum wax, or other synthetic ingredients. And even so-called natural chewing gums like Glee Gum, which is made with natural chicle (rubber from a tree), combine the natural materials with synthetic polymers to make its gum base. The only way to know for sure whether your chewing gum is plastic or not is to contact the company and find out the exact ingredients used to make the gum base. If they won't tell you, don't chew it. Recently, a new company has come on the chewing gum scene offering natural, plastic-free gum in plastic-free packaging. Simply Gum contains only organic dried cane juice, all natural chicle, organic vegetable glycerin, organic sunflower lecithin, organic rice flour, and natural flavors of cinnamon, maple, fennel licorice, ginger, mint, or coffee.

Baby Food

A lot of moms are switching to homemade baby food to avoid chemicals in processed foods, as well as BPA from glass jar lids. The Momtastic website provides a wealth of information and recipes for making your own baby food. Many moms freeze small portions of food in ice cube trays and store it in plastic-free containers. I'll talk more about food storage containers and plastic-free ice trays in chapter 9.

Pet Food

While Michael and I don't have a human baby, we do have two furry black feline "kids" that we love ridiculously. In fact, one of them lies at my feet as I write this book and competes with the computer for my attention. Concerned about the quality of ingredients in processed pet food and irritated at all the packaging waste from bags and BPA-lined cans, we decided to try making our own cat food, which we've been doing for several years. And when I say "we," I mean Michael. I'm the one who came up with the idea, and he's the one who follows through. See how that works?

> "Making your own baby food doesn't really take that much time, even for a domestically challenged person like myself! When my kids were babies, I used the book *Super Baby Food* by Ruth Yaron (F. J. Roberts Publishing Company) as a guide and made batches of food for the freezer." —Katy Farber, www.non-toxickids.net, Burlington, Vermont

While there are many recipes online for making pet food from various raw or cooked ingredients, we went with the BalanceIT system, which was developed by veterinarians at the UC Davis Veterinary School, and which provides complete nutrition for our cats, so we don't have to worry whether their fussy feline bodies are getting all the nutrients they need. The system is not completely plastic-free, but it involves much less packaging than commercial alternatives.

Here's how the system works: On the BalanceIT site, pet owners enter the type of animal (dog or cat); select a type of meat and a type of carbohydrate they want to use; and provide the gender, age, weight, and physical condition of the pet. The system creates a customized recipe for the animal, which includes BalanceIT supplement powder, a formulation that includes all of the vitamins and minerals the animal needs to stay healthy. Our homemade cat food contains ground poultry, which Michael buys from the butcher counter in our own stainless steel pot; baked yams, which come from the farmers market or produce aisle; and supplement powder. The powder comes in a plastic bottle with a plastic scoop, but since one bottle lasts two months, this system generates much less packaging waste than commercial alternatives. And it gives Michael joy to provide for our kitties in this way. It does. Just ask him.

OH, THE IRONY: ORGANIC FOOD IN PLASTIC PACKAGING

Switching to plastic-free eating meant reducing the chemicals from plastics going into my food. And that, in turn, lead me to research more about other types of chemicals in our food supply: pesticides, hormones, antibiotics, preservatives, artificial colors and sweeteners and flavors. I committed to eating organic food as much as possible to avoid those chemicals. So I was mystified by the irony that so many organic food companies package their products in plastic. How, after taking care to eliminate hormone-disrupting chemicals while growing the food, could they then package the food in a material that could potentially leach chemicals right back in? I'm guessing that food producers simply don't know about all the chemicals in their plastic containers and wrappers and believe they are safe based on assurances from plastic producers and the FDA. According to a representative of the Stonyfield Farm organic yogurt company, the plastics industry is very secretive and doesn't divulge its proprietary formulas even to the companies that use its products to package their foods.

Foods wrapped in plastic should not be certified organic, no matter how they are grown. Do you agree?

Personal Worksheet: Grocery Substitutions

Use this sheet to list the top 20 food items you buy regularly that come packaged in plastic. As you discover a plastic-free solution, mark it on the list. Solutions could include a different brand, store, recipe, alternate product, or even doing without. Use the list to make one change every one–two weeks, depending on availability of plastic-free products in your area.

Type of Grocery Item	Brand I Buy Now	Less-Plastic Alternative

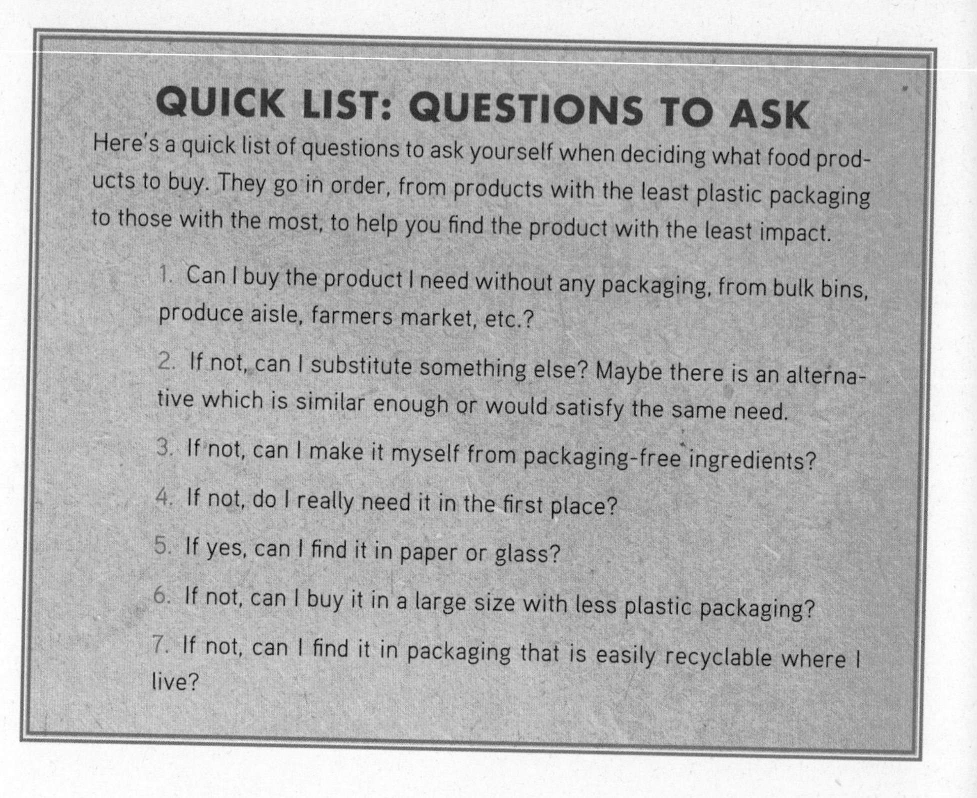

QUICK LIST: QUESTIONS TO ASK

Here's a quick list of questions to ask yourself when deciding what food products to buy. They go in order, from products with the least plastic packaging to those with the most, to help you find the product with the least impact.

1. Can I buy the product I need without any packaging, from bulk bins, produce aisle, farmers market, etc.?
2. If not, can I substitute something else? Maybe there is an alternative which is similar enough or would satisfy the same need.
3. If not, can I make it myself from packaging-free ingredients?
4. If not, do I really need it in the first place?
5. If yes, can I find it in paper or glass?
6. If not, can I buy it in a large size with less plastic packaging?
7. If not, can I find it in packaging that is easily recyclable where I live?

Community Action: Plastic-Free Farmers Markets & Grocery Stores

About two months into my plastic-free project, after lying awake for hours ruminating about what lay ahead, I heaved myself out of bed at 6 AM on a Sunday morning; strapped a folding table and chair to a mini hand truck, along with a basket of Green Sangha's "Don't Think About A Plastic Bag" flyers and my knitted plastic bag fish Tina; and lugged it all half a mile down the street to the Temescal Farmers Market in North Oakland to spend the morning educating the customers about plastic bags. At that time, farmers markets were just waking up to all the plastic waste they were generating from produce bags and other plastic containers, and many markets in the Bay Area wanted to change. But they also didn't want to alienate their customers, who had gotten used to the convenience of the handy rolls of free plastic bags provided at each produce stand. So, I decided it was up to me to explain to all the people why plastic bags were bad.

Standing beside my table in the "free speech" area of the farmers market near the entrance to the parking lot where most people entered and exited, I called out to each person who passed by, "Can I give you some information about plastic?" As I expected, the reactions were mixed: some people took flyers politely; others outright refused or looked away; and a few started to walk away until they heard the word "plastic" and then actually turned around and came back to hear what I had to say, clearly relieved that I didn't seem to want their money or their immortal souls. I started the day a little apprehensive and nervous, but as the morning wore on and I got into a groove, I really started to enjoy myself. One woman stopped, realized she had left her bags in the car, and said with a sigh, "Just because of you, I'm going to go back and get them." Half an hour later, she proudly showed me her plastic-free purchases on her way out.

It was an exhausting and exhilarating morning, and while it was sometimes hard to get up the courage to talk to strangers, I pushed myself because I knew I had to go beyond my own personal changes if I was going to make a dent in the plastic problem. I had to speak up. I didn't have a choice. And because of so many other people who have made the effort to write letters and speak out, the times they are a-changin', at least as far as Bay Area Farmers Markets are concerned.

Plastic-Free Farmers Markets

In 2009, the Berkeley Farmers Market became the first in the nation to ban plastic bags and packaging. The vendors there offer compostable bags, for which they charge 25 cents to encourage people to bring their own bags instead of taking the disposables. The following month, both the Fairfax and San Francisco farmers markets went plastic-free. My own Temescal farmers market kicked the plastic habit in January of 2010, with some vendors providing reusable baskets for customers to use before emptying their purchases into their own bags or offering compostable bags for a fee. And the movement is spreading.

Grocery Stores Say No to Plastic

Changes are happening in grocery stores as well. For example, with the help and support of Green Sangha's Rethinking Plastics campaign, the Good Earth natural food store in Fairfax, California, has significantly reduced its waste from plastic produce, bulk, and fish and poultry bags. In the first year of its anti-plastic campaign, the store went from using 3.2 tons of plastic per year to 1.88 tons the following year. How did they manage that? By planning out a vigorous educational campaign to win their customers' support. Here are some of the steps they took:

- Built a display of numerous items a person could incorporate into their life to reduce plastic use: Klean Kanteens, glass containers, organic cotton bulk/produce bags. The display included instructions for how someone could use these items in their household and while shopping.
- Posted signs explaining to customers why a bag isn't necessary for most produce items. "Nature has amazing packaging."
- Published articles about plastic in their monthly newsletter, explaining the problems and what the store was doing to reduce plastic use beyond eliminating grocery bags.
- Placed signs in the parking lot reminding people to bring in their reusable bags.
- Removed most plastic bag options inside the store, including most rolls of plastic bags in the produce department, all plastic bags in the bulk department, and all plastic bags in the fish and poultry department, only using them at customers' request. They left only two rolls of very thin

plastic bags in the produce department and posted signs next to them with encouraging facts about saving sea turtles.

- Provided paper bags and offered reusable cloth bags for sale. Hopefully customers will get the message that paper bags have an environmental impact as well, and that, as with reusable grocery bags, the best solution is bringing their own.

After eliminating plastic shopping bags, many businesses would like to go further and eliminate other forms of plastic packaging. But the only way they'll take these steps is if they know their customers want it. It's up to us to write letters, send emails, speak to market managers, and organize our communities to ask for what we want.

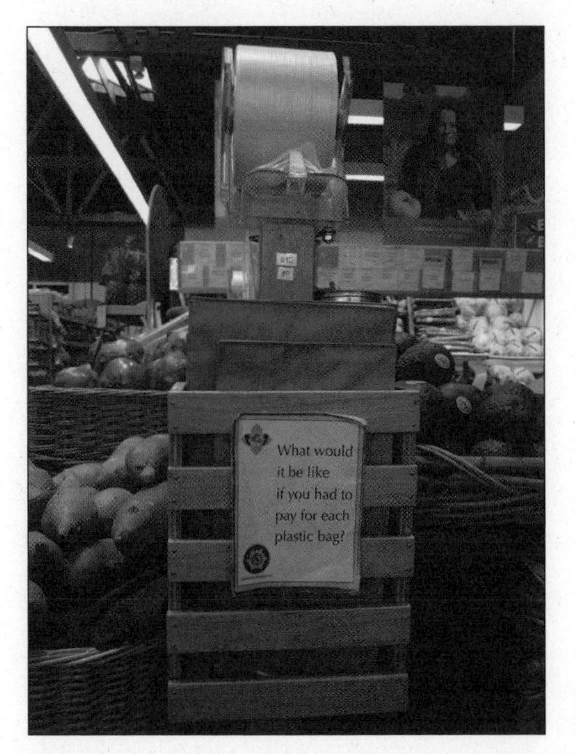

Good Earth anti-plastic bag campaign

Action Items Checklist

(Choose the steps that feel right to you. Then, as an experiment, challenge yourself to do one thing that feels a little more difficult. Only you know what that one thing is.)

- ☐ Obtain reusable produce/bulk bags/containers.
- ☐ Find locations and times of area farmers markets.
- ☐ Visit farmers market to learn what's in season without plastic packaging.
- ☐ Talk to the farmers market management and vendors about eliminating plastic produce bags and other plastic packaging.
- ☐ Locate stores selling foods in bulk.
- ☐ Visit bulk bins to learn what you can substitute for packaged foods.
- ☐ Try making your own condiment or other processed food. (Do the processing yourself!)
- ☐ Use Grocery Substitutions Worksheet to make a plan to eliminate one plastic-packaged item every 1-2 weeks.
- ☐ Talk to store manager about adding more bulk items.
- ☐ Talk to a store manager about getting rid of plastic produce bags and other types of plastic packaging.
- ☐ Write to a food company about switching to plastic-free and BPA-free packaging.
- ☐ Write to organics organizations and ask what we can do to make plastic packaging a factor in organic certification.

Chapter 7: Personal Care & Household Cleaning (When Lazy = Green)

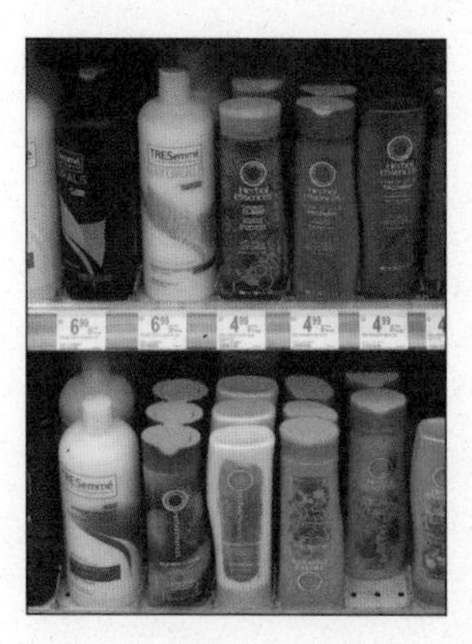

If figuring out ways to feed myself without plastic challenged my homemaking skills—learning to cook and garden—figuring out how to replace all the plastic related to keeping my body and home clean and well-maintained—bottles of shampoo, conditioner, lotion, and mouthwash; toothpaste tubes; toothbrushes; dental floss; deodorant sticks; razors; cans of shave gel and air freshener with plastic tops; containers for lipstick, blush, powder, eye shadow, and mascara; plastic wrappers for toilet paper, paper towels, facial tissue, and cotton balls; liquid soap pumps; feminine hygiene products; bottles of laundry detergent, cleaning sprays, and dish soap; synthetic sponges and scrubbers—really tested my ingenuity. I experimented a lot. In fact, I had to learn a little chemistry.

Six months into my plastic-free experiment, we used up the last of our liquid hand soap. The easiest solution would have been to just switch to bar soap, which is what we use nowadays. But back then, I figured that it shouldn't be hard to make liquid soap out of a bar of soap. Just add water and let it dissolve, right? So I bought a 1,000 gram (2.2 lbs.) block of unpackaged olive oil soap from a local natural products store, put the entire block in a stock pot on the stove, filled the pot with water, and started stirring. And stirring. And stirring. Wanting to save time, I hadn't bothered with grating the soap beforehand, so what should have taken a few minutes ended up taking a few hours. Hours of gas stove energy. Hours of stirring.

After the soap block was fully dissolved, I let the pot cool. The next morning, I checked on my creation, only to find a huge stock pot of solid soap! Guessing I should have used more water, I dug out half of the solid soap and set it aside, then topped off the pot and started the process again. Heat on. Stirring and stirring. The next day, I checked on my creation and found . . . another pot full of solid soap! Feeling a bit like Bill Murray in the movie ***Groundhog Day***, I repeated the process over the next few days, until finally I had a

Plastic-free personal care items

substance that seemed somewhat pourable, as well as several additional bowls of soap in various stages of solidity.

Now the problem with my "liquid soap" was that it was full of little crystals that hadn't dissolved. So I poured it through a metal strainer. That got out the larger bits, but the soap still wasn't smooth. So I poured it again through cheese cloth. Finally, I ended up with a slimy, stringy, gloppy, green mess that had consumed massive amounts of time and energy and, in the end, was kind of disgusting. It would have been great at a Halloween party, maybe, but wasn't so nice for hand or dish washing.

It turns out that liquid soap is not simply dissolved bar soap but is in fact chemically different. (Bar soap is made with ***sodium*** hydroxide; whereas, liquid soap is made with ***potassium*** hydroxide, which makes for a more liquid consistency.) However, you can make a passable version of liquid soap from a bar of soap if you do it the right way. I've included instructions in this chapter. The point to this story is that learning to live with less plastic requires not only a willingness to experiment but also to fail, to laugh at ourselves, and to try again. Over and over. Until we get it finally right, and move on to the next adventure.

Over the last five years, I've discovered many solutions for plastic-free personal care and cleaning, which I'll share with you in the next few pages. But this list is by no means complete. Maybe you've discovered some solutions that I haven't thought of, or maybe this list will inspire your own spirit of innovation. I'm counting on it.

What's in the Bottle?

Before I talk about ways to avoid plastic products and packaging, I want to say a quick word about all the chemicals we are exposed to every day from our personal care and cleaning products. The Environmental Working Group (EWG) estimates that the average woman uses twelve different personal care products each day, and the average teen uses seventeen, not to mention all the creams, soaps, and lotions used on babies.[101] The chemicals we put on our bodies are absorbed through our skin and can enter our bloodstream, and sadly, many of these chemicals—including carcinogens, pesticides, reproductive toxins, endocrine disruptors, plasticizers, degreasers, and surfactants—are harmful.[102] When considering which products to buy, it's important to be mindful of both the packaging ***and*** the chemicals in the product itself. For this reason, I have limited the solutions in this chapter to those products that seem to be the healthiest both inside and out. If a product has less plastic packaging but is full of harmful chemicals, I won't promote it. Or I will note the drawbacks in my description.

For more information about chemicals in cosmetics, read Stacy Malkan's book, *Not Just a Pretty Face: The Ugly Side of the Beauty Industry* (New Society Publishers, 2007). Visit the Campaign for Safe Cosmetics website to learn what you can do to push for safe cosmetics legislation. For help choosing safer products now, visit EWG's Skin Deep Cosmetics Database, which lists over 70,000 products from over 2,000 different brands. Be aware, however, that while Skin Deep lets you know what's in the product, it doesn't tell you anything about the packaging on the outside.

A note about LUSH brand products: In the first edition of this book, I mentioned LUSH products multiple times because so many of them come "naked" without any packaging. Since that edition came out, I have rethought including LUSH products in this book. While the lack of packaging is great, it's the ingredients inside some LUSH products that concern me—ingredients such as synthetic fragrances, colors, and preservatives. However, when I compare the ingredients in LUSH's "naked" solid products to those in conventional products, I still think they come out ahead. (Parabens, for example, are only found in products packaged in plastic containers.) So, while I can't give LUSH an unqualified endorsement, I will mention their products throughout this chapter, and any time I do, I'll also let you know about ingredients of concern, so you the reader can decide if they are right for you.

Tips for Cleaning Up Good . . . without Plastic

Bulk Buying, Revisited

Some co-ops and natural food stores sell personal care and cleaning products in bulk and allow you to bring your own bottles to fill. I've seen shampoos, conditioners, liquid soaps, lotions, laundry detergent, dish soap, and other products sold this way. Dr. Bronner's liquid soaps are a good choice because not only is the soap all natural, fair trade, and organic, but the plastic jug is itself made from 100 percent postconsumer recycled plastic. However, it,

LESS-PLASTIC HEROES: Bettina Limaco and Marco Pietschmann of Green11

San Francisco is full of shops selling specialty cosmetics, lotions, creams, soaps, sprays, and all manner of formulations that promise to keep our outsides beautiful. And nearly all of these products are sold in plastic packaging. But in 2009, Bettina Limaco and her husband Marco Pietschmann opened Green11, a new kind of shop where all the products are refillable. Like bulk food store in.gredients, Green11's mission is to get customers to focus on the product over the packaging. Since a lot of people come home and transfer their liquid soaps and lotions into nicer bottles anyway, it makes sense to let them fill their own containers from the start and eliminate the throw-away plastic. I dropped by the store in the summer of 2011 to have a look around and ask Bettina how she and her husband got started.

Bettina explained that as an accountant with a background in consumer products, she had learned firsthand what the cost of that plastic bottle is, and it just didn't make sense to spend so much effort and money creating packaging that not only hurt the environment, but didn't even add value to the consumer. What's more, she and her husband had witnessed the environmental effects of plastic packaging firsthand while scuba diving in Egypt for seven weeks. On some of the most remote beaches on the planet, they saw plastic everywhere they looked, including shampoo bottles and laundry soap containers.

Once home from their long trip, the couple set out to create a different kind of store, offering healthy personal care products (Marco uses EWG's Skin Deep Cosmetics Database extensively to research products and make decisions about what to carry) without extra packaging. The store also offers many cleaning products, such as laundry and dishwasher detergents and dish soaps. In fact, Bettina says that while many of the personal care products come in big plastic jugs, the biggest plastic savings is with laundry soaps and cleaners, some of which come in huge 55-gallon drums.

I asked Bettina what happens to the large plastic bulk containers when they are empty. She said that while it's true that most of the product companies will not take back their containers to refill, it's really important to her to see that the containers don't go to waste. She posts ads in the Craigslist free section to give away drums and pails and has compiled an email list of people who want to take them off her hands to repurpose for activities such as collecting rainwater or mixing paint.

While I was visiting the store, a young woman came in with an empty Method liquid soap bottle to refill. Green11 doesn't carry Method products, so she filled it with a different brand, explaining that while she knows she can put the Method bottle in her recycle bin, doing that feels wasteful compared to refilling it. While she likes Method products, being able to reduce waste is more important to her than brand loyalty. Let's hope that with increasing numbers of us voting with our wallets to reduce excess plastic packaging, companies will finally get the message.

like almost all of the other large plastic jugs that contain bulk products, doesn't get sent back to the company for refilling. Instead, stores and co-ops are left to recycle them as best they can. So, as with foods in bulk bins that come packaged in big plastic sacks, buying bulk personal care products in plastic jugs is a "less-plastic" solution rather than completely plastic-free. It's a much better choice than buying new bottles each time, but there are ways to reduce our plastic consumption even more.

Don't Pay for Water

Browsing the Internet tonight for the ingredients in popular liquid laundry detergents, I see that water is always first on the list. We can eliminate the need for plastic containers altogether by choosing solid or powdered products instead of liquid. Soap, shampoo, conditioner, and deodorant bars; solid lotions; powdered laundry and dishwasher detergents; and toothpaste tablets and powders are just a few examples. Why pay for water when we can add it at home from our own taps? I'll list specific brands and recipes I've found in the next sections.

An easy first step in de-plasticking your personal care products is to switch from liquid soap to bar soap. Several brands (Sappo Hill, for example) come completely unpackaged. People sometimes worry that sharing a bar of soap is less sanitary than sharing a bottle of liquid soap. But think about it: the bar soap gets rinsed off every time you use it. The plastic pump? Not so much.

Still, there are times when liquid soap comes in handy. For example, many homemade cleaner recipes require a few drops of liquid soap, as do recipes for natural pesticides, a fact I learned when I began gardening. Here, then, is a recipe for making a passable kind of liquid soap from a bar of soap. It's not as concentrated or smooth as real liquid soap, but it works.

"The best way to exfoliate your skin is to simply use a fresh clean washcloth. Forget nylon shower puffs and exfoliating microbeads— they are plastic! Washcloths work better anyway." —Juli Borst, classical singer and urban treehugger, www.plasticlessnyc.blogspot.com, New York City, New York

How to Make Liquid Soap from Solid Soap

Ingredients and tools:

- 4 ounces (120 g) of solid natural soap (I used a plastic-free bar of Sappo Hill soap)
- 1 gallon (4 L) of water
- Cheese grater
- Stock pot
- Electric mixer, immersion blender, or wire whisk (not essential, but it helps.)

1. Grate the solid soap. This is an important step to a well-mixed result. You can also save up soap pieces from used bars of soap.
2. Bring water to a boil in a steel pot and remove from stove. Pour grated soap into pot and stir until dissolved.
3. Let sit over night. The next day, the soap will have firmed up into a thick gel.
4. Blend with hand mixer, immersion blender, or wire whisk. This step is not necessary but ensures the soap is well-mixed and also adds air to fluff it up and make it less slimy.
5. Bottle soap in a glass jug or repurposed pump container.

Plastic-Free Skin Care

Let me just say up front that I'm lazy. I wash my face with plain water while I'm in the shower. I get a pedicure only once a year before I attend the annual BlogHer conference for women bloggers. I put on makeup when I think someone might take a picture of me, which isn't often. And moisturizer? I must not need it because I never use it. For years, I thought that my laziness was simply a sign that I was, you know, lazy. But now I know it's a way to be green. After all, the fewer products we use, the fewer resources we consume. So as you read the solutions in the next few pages, keep in mind that keeping your personal care regimen as simple as possible with as few ingredients as possible is not only healthier and more eco-friendly, it saves time and money too.

Don't Flush Plastic Down the Drain. Many exfoliating scrubs (and other personal care products, including toothpaste) contain tiny microbeads made from polyethylene plastic, plastic that is meant to be rinsed down the drain. Water treatment facilities are not designed to filter out such tiny particles, so they enter our waterways and the bodies of aquatic creatures. Check the ingredients list of any scrubs you are considering buying to make sure they don't contain "polyethylene" or "microbeads." Beat the Microbead is an international campaign spearheaded by the Plastic Soup Foundation in the Netherlands. Visit their website for lists of products that contain microbeads and a mobile app that you can use while shopping. The site also lists nonprofit organizations throughout the world (like 5 Gyres mentioned in Chapter 1) that are part of the campaign. Click the organization links to learn what you can do in your country.

Make Your Own Scrubs and Facial Products. Plain baking soda is a great exfoliant/facial cleanser. It's probably the cheapest and simplest as well (aside from plain water). Just make a paste in your hand with a little water and scrub away. Other ingredients for skin cleansers are sea salt, oatmeal, finely ground almonds, flax seed meal, ground lentils, brown rice flour, coffee grounds, citrus fruit peels and mashed fruits, honey, and sugar, most of which you can probably find in bulk. Search the Internet for recipes using these ingredients. Or get a copy of the book *Better Basics for the*

Home, by Annie Berthold-Bond (Three Rivers Press), which contains a wealth of ideas for DIY personal care products without toxic chemicals.

Make Your Own Clay Masks. Instead of purchasing expensive clay masks in tubes and jars, see if your bulk foods store sells bentonite, kaolin, and other food-grade powdered clays in bulk. The few times a year my pores need extra attention, I mix up some bentonite clay with apple cider vinegar, which happens to come in a glass bottle. Blackheads beware!

Moisturize with Olive or Coconut Oil. Many people swear by plain olive oil or coconut oil. You can search the Internet for natural moisturizer recipes using lots of different edible ingredients, but to me, the best option is to find a solution using the fewest ingredients possible. Then again, I'm lazy. Like I said, I don't moisturize at all. Your mileage may vary.

Choose Lotions and Lip Balms in Plastic-Free Containers. Organic Essence packages its organic hand and body creams in compostable cardboard containers and its lip balms in ingenious cardboard tubes that squeeze from the end so you don't have to touch the product with your fingers. You can also find solid lotions in metal containers or even packaging-free. And many Etsy sellers create solid lotions and lip balms packaged in metal tins or cardboard tubes instead of plastic tubes. One of my favorites is the luxurious vegan lip balm from Juniperseed Mercantile in its oversized cardboard tube. Or look for recipes for making your own lotions and creams from bulk oils and vegetable waxes.

Soothe Diaper Rash with Natural Products Packaged in Glass or Metal. There's no need to resort to plastic-packaged baby lotions and diaper rash creams. Waxelene is a natural alternative to petroleum jelly and comes in a glass jar with a metal lid, as does the Diaper Balm from Taylor's Pure and Natural. MadeOn Lotion offers a rash cream that contains only three ingredients—coconut oil, zinc oxide, and beeswax—and comes in a reusable metal tin. Other companies to try are The Rex Apothecary and Aquarian Bath.

Choose Plastic-Free Sun Protection

First, let me say that when it comes to sunscreen, it's important not only to consider the packaging, but also to avoid many toxic chemicals found in conventional sun products. It appears that the safest ingredient for non-toxic sun protection is non-nano zinc oxide, which is the active ingredient in all three of the following products: The Rex Apothecary offers a vegan sunscreen that comes in a metal tin; Taylor House's vegan BALM Baby! is rated SPF30, comes in a glass jar with a metal lid, and smells mostly like lavender; and Avasol is a solid sunscreen that comes in a compostable cardboard tube and smells like cinnamon. It's available in two different shades—tan and dark—to avoid the zinc oxide pallor. Avasol is not vegan, as it contains beeswax, propolis, and emu oil. I tried both BALM Baby! and Avasol under the burning sun of the Black Rock Desert at the Burning Man art festival last year and both worked great on my fair skin.

These natural, non-toxic sun products are much more expensive than what you might buy at the local drug store. So to reduce the amount of product you need to use, consider other practical ways to reduce sun exposure. Bare less skin, for example. Wear a hat, longer sleeves, and capri pants instead of shorts, perhaps. And stay out of the sun during the middle of the day when the rays are the most direct. You might even consider carrying a parasol. I'm not kidding. They seem to be making a comeback.

Shaving/Waxing

The simplest answer here is just to forego shaving or waxing and let our hair grow as it wants. But for those of us not willing to sport hairy chins, legs, and armpits, there are solutions.

Switch to a Metal Safety Razor or Straight Razor. When I first learned to shave my legs back in the 1970s, I used my dad's metal safety razor because it was what we happened to have in the house. It worked fine. But you can believe when Personal Touch came on the scene with its advertising targeted at women, I leaped at the chance to have my own pretty "tortoise shell" razor. Of course, no tortoises were harmed (thank goodness) in the making of the Personal Touch razor because it was made entirely from plastic. But at least the handle was reusable. Then, Bic came out with completely disposable razors. As teenagers, my sisters and I would leave piles of half-rusted throw-away razors all over the bathroom for our imaginary maid to clean up. But now, as a responsible adult, I've gone back to the metal safety razor I started out with. Well, not the exact same one. I found mine in, of all places, a local antique shop! While you can buy brand new ones at Life Without Plastic, The Art of Shaving and Classic Shaving, do the planet a favor by giving a home to a secondhand vintage safety razor or straight razor from a local antique store or eBay. Make sure you request to have your razor shipped without plastic packaging!

Buy Safety Razor Blades without Plastic. While safety razor blades are plastic-free, the packaging usually is not. Personna brand blades can be purchased from eBay sellers in cardboard boxes of 100 to 1,000 blades. Genuine Personna double-edge razor blades are made in Israel and last a very long time. In fact, I'm pretty sure I'll still be working on my first box of blades until the day I die.

Opt for Recycled Plastic. While using a metal razor is the plastic-free option, some people are hesitant to go there. The reduced plastic choice is the recycled plastic razor from Preserve, which can be returned to the company through Preserve's Gimme5 program for recycling. But seriously, try a secondhand metal razor. It's not scary. And if you don't like it, you can always sell it back through eBay. Classic shavers are growing in popularity these days.

Use Shave Soap Instead of Canned Shave Cream or Gel. Back in the day, men bought round cakes of shave soap, which they kept in a cup and lathered on their faces with a brush. Shave soaps still exist. In fact, that's what Michael uses every day. Brands include Simmons Natural Bodycare, Williams Mug, Soaptopia, and The Rex Apothecary, and many Etsy sellers offer handmade shave soaps. Shave brushes can generally be found wherever metal razors and shave soaps are sold. But while Michael loves lathering up with his brush and mug, I just use a plain bath bar. Simple. Lazy. Green.

Make Your Own Sugar "Wax." Once a year, I get my crazy eyebrows professionally waxed, and for a week or so, my face looks human again. But of course, most hair removal wax is petroleum-based and comes in plastic packaging. I discovered recently how easy it is to make "wax" from sugar and honey. What's more, it doesn't hurt as much! On page 234 is a recipe that makes a small amount of wax, just enough for eyebrows. Search Google for recipes for larger quantities, and search YouTube for videos demonstrating exactly how to do it.

Get Out the Tweezers. Tweezing requires a lot less effort than making up a batch of wax. I prefer getting the pain over with all at once, but if you can stand the slow pluck, pluck, pluck, go for it.

Deodorant

Everyone's body and body chemistry are different. That's the main factor to keep in mind when looking for a plastic-free deodorant solution. What works great for me may leave you smelling like a litter box, and what makes me itch like crazy may be the best thing you ever tried. So here are some ideas for plastic-free deodorants. Hopefully something on this list will work for you, or at least get your imagination going.

Note: the solutions here are deodorants, not antiperspirants. They are not meant to eliminate wetness, only odor.

LUSH "Naked" Deodorant Bars LUSH sells bars of deodorant in naked chunks, rather than in containers. If you like the smell of LUSH deodorants and can tolerate the ingredients (the Aromaco bar does contain propylene glycol), this might be a good alternative.

Use baking soda as deodorant.

Rex Apothecary Deodorant Bars have a tiny bit more packaging than LUSH but contain no propylene

Homemade Sugar Wax Recipe

Ingredients:

- 2 teaspoons (8 g/10 mL) brown sugar (bulk bin)
- 1 teaspoons (5 mL) honey (bulk bin or glass jar)
- 1 teaspoons (5 mL) water (tap)
- Cloth strips (Cut up old sheets, hankies, pajamas, or any other type of non-stretchy fabric you have on hand. T-shirt material and other knits are not recommended, as they're too stretchy.)

1. Combine ingredients in a microwavable container. I used a Pyrex measuring cup. You could also use the stove instead of microwave, but you'd need a very tiny pan.
2. Cook mixture in the microwave (or on the stove) until it bubbles and turns brown. This part can be challenging. If you don't cook the mixture long enough, it stays too soft and sticky to work. If you cook it too long, it turns into hard candy. For me, cooking for 30 or 35 seconds is about right.
3. Let cool.
4. Apply wax to only the part of the brow you want to remove. Be careful! This stuff really works. I learned this lesson the hard way and have the pictures to prove it.
5. Apply cloth strip to waxed eyebrow. Press and smooth in the direction of the hair growth. (Cloth strips can be washed and reused because sugar dissolves and washes away cleanly, as opposed to real wax which ruins fabrics.)
6. The moment of truth. When you're ready, rip the cloth strip off of your eyebrow in the opposite direction of the hair growth. And . . . voila!

glycol or other synthetic ingredients. They come packaged in a compostable, corrugated cardboard wrapper lined with unbleached soy wax paper.

Baking Soda with Tea Tree Oil You know how baking soda neutralizes odors in your refrigerator? It works that way on your body as well. I use plain baking soda with a few drops of tea tree essential oil. I keep the powder in a repurposed tea tin in a dresser drawer and apply it with a reusable cotton round (more on cotton rounds later in this chapter). For eliminating odor, baking soda works better for me than any commercial deodorant ever has. And the tea tree oil adds some extra antifungal and antibacterial protection. Lavender is another antibacterial choice.

Baking Soda and Cornstarch Some people are sensitive to baking soda and start to itch after using it for a while. Mixing it with cornstarch can help alleviate that problem. But note that reducing the percentage of baking soda reduces the odor protection. Add some tea tree or other essential oils as desired.

Baking Soda, Cornstarch, and Coconut Oil Several Internet sites recommend mixing baking soda or baking soda and cornstarch with coconut oil. Coconut oil is said to have some odor-fighting properties itself. I personally didn't find this formula to work as well as plain baking soda, but many people love it. Add essential oils as desired.

Vinegar and Water Spray Some people swear by a spritz of vinegar and water under the arms. I tried it. Made me smell like a salad and didn't fight odor. But like I said, we're all different.

Alcohol and Water Spray Alcohol kills the bacteria that cause odor. But rubbing alcohol comes in plastic bottles and contains chemicals which make it hazardous if swallowed. Instead, for personal care and cleaning purposes, try vodka in a glass bottle or better yet, Everclear, which is 75 percent (in some states, up to 95 percent) alcohol. A very little goes a long way. However, this alternative may not be appropriate for people with drinking problems, as you'll see in the next chapter.

Lavender Essential Oil Either dilute with water and spray on or dab no more than a drop or two under each pit. One of my blog readers uses this method daily and loves it.

What About the Crystal? Deodorant crystals are made from potassium or ammonium alum, mineral salts that are thought to be harmless. Many natural living proponents love them. But brands sold in stores come in a lot of plastic packaging.

Deodorants in Metal Tins or Glass Jars Chagrin Valley offers a natural deodorant in a metal tin. So does Aquarian Bath. Taylor House's ThincSkin deodorant comes in a glass jar with a metal lid. Check Etsy.com for sellers offering deodorant formulas in recyclable/reusable tins instead of plastic.

Recyclable Plastic Deodorant Tubes If none of the above solutions work for you, and you can't come up with a plastic-free alternative on your own, consider choosing a deodorant from Tom's of Maine. Its plastic deodorant tubes can be downcycled via its program through Terracycle.

Aquarian Bath solid shampoo bar

Hair Care

Reducing plastic packaging doesn't mean sacrificing healthy hair. There are quite a few plastic-free options these days.

Solid Shampoo and Conditioner Bars Just as there are solid soaps, lotions, and deodorants, there are solid shampoo bars and conditioners, some of which come without any packaging at all. Natural brands include Aquarian Bath, Chagrin Valley, Beauty and the Bees, and

J.R. Liggett (which can often be found on the shelves of natural food stores). LUSH and Basin also sell shampoo and conditioner bars without packaging, but they contain synthetic ingredients. Many Etsy sellers offer handmade shampoo and conditioner bars as well. How do you use them? Either rub the bar directly on your scalp or rub some into your hands and scrub through your hair and scalp. Rinse as usual. No need for any plastic bottle.

"No 'Poo" Hair Cleansing Regime If you haven't heard of the "No 'Poo" method of hair cleansing, you'll need to keep an open mind as you read about it. I have been washing my hair this way for the past seven years, and it works great. My hair is very healthy, albeit turning a little gray. Here's how it works:

1. Wash hair with a solution of baking soda and water. One tablespoon of baking soda per cup of water. Pour it on scalp and scrub with fingers in shower.
2. Rinse hair with a solution of apple cider vinegar (ACV) and water. Same proportions: one tablespoon of ACV per cup of water. I add a few drops of rosemary essential oil for a nicer smell and antidandruff benefits. Work it through hair and make sure to saturate the ends. The ACV conditions hair and makes it soft and manageable. I mix up both solutions and keep them in repurposed plastic sports bottles in the shower.
3. Rinse with water if desired (this step is optional) and comb out. The vinegar smell will dissipate very quickly. Don't believe me? Ask your friends to sniff your hair when it's dry. I guarantee they won't guess what "conditioner" you used.

That's pretty much it. If you've been washing your hair with commercial products for a while, it may take several weeks for your scalp to get used to the new system. Some people experience a period of greasy hair, as the scalp continues to overproduce oil. But with patience, you should notice your hair settling down and looking healthy and shiny. In fact, after a while, you may notice that you don't need to wash your hair as often as you did when using shampoo. For help with any issues that may arise, check out the No 'Poo Discussion Forum (www.no-poo.livejournal.com), whose tag line is "No, we're not constipated. Yeah, we get that a lot."

Deep Condition with Olive Oil, Coconut Oil, or Mashed Avocado There are quite a few recipes on the Internet for conditioning treatments using these ingredients. Basically, you just massage the oil through your hair, cover with a warm towel, leave it on for 30 minutes, and rinse.

Natural Pomades in Glass Jars or Metal Tins Pomades are a great choice for controlling frizzies. My favorite product used to be called just that: Product. It comes in a glass jar. Unfortunately, the jar has a plastic lid, but a little Product lasts a very long time if used sparingly. My new favorite frizz control product is MadeOn Second Life hair butter, which comes in a metal tin and contains only four ingredients: shea butter, coconut oil, beeswax, and orange essential oil. You can also find many different handmade pomades in metal tins and glass jars on Etsy.com. Be sure to specify no plastic shipping materials when ordering.

Henna Hair Color LUSH sells solid henna hair dye bars in various colors with no plastic packaging. The bars are made with cocoa butter and essential oils. You may also be able to find henna powder in bulk at your local co-op or natural foods store. I've tried dying my hair with both henna powder and LUSH henna bars, and both seem to work equally well. I've even had success touching up my roots by using an old toothbrush to apply the henna just where I want it. Keep in mind that using henna can be a messy task. Wear gloves (Yes, you can find them plastic-free!) if you don't want to end up with black or red fingernails, and designate a few old towels as henna towels because they will get stained. Instructions for using henna will usually advise you to cover your head with a plastic bag while waiting for the henna color to set, but old towels work fine for me.

Plastic-Free Wooden Hairbrush and/or Comb Don't go out and replace your perfectly good plastic brush and comb. But if your old ones wear out and you need new ones, check out the offerings at Life Without Plastic. My favorite hairbrush has wooden bristles set in a natural rubber base and ships with no plastic packaging.

Dental Care

Brush with Bamboo When it comes to toothbrushes, none are perfect, but some come closer than others. While there are several brands with wooden or bamboo handles, the bristles are either made from nylon—a type of plastic—or pig bristles, a by-product of the meat industry. However, there is one guy working to create the most eco-friendly toothbrush possible. His name is Rohit Kumar, and his company is Brush with Bamboo.

Since its inception, the Brush with Bamboo toothbrush and packaging have gone through several revisions. At first, the handle had a nice curved shape that fit well in your hand. I liked it. But in order to create that shape, several pieces of bamboo had to be glued together, and Ro couldn't be assured that the adhesive was completely non-toxic and biodegradable, so he scrapped that design and switched to a handle made from a solid piece of bamboo. It's still curved, but not as much as the original, which to me is a decent trade off in favor of sustainability.

For the bristles, Ro chose nylon because he didn't want to use an animal product. But he's working on developing bristles from a compostable American-made polymer derived from castor beans. Perhaps by the time you read this book, that new version will be available.

And Ro doesn't only care about the toothbrush itself but also the packaging. The toothbrush comes in a cardboard box that is held together with tabs—no glue or plastic window. Inside the box, the toothbrush is wrapped in a compostable bag made of corn-based PLA plastic. But because there are drawbacks to plastics made of GMO corn (as I discussed in Chapter 5), his new version will be packaged in a compostable NatureFlex wood cellulose wrapper instead.

There are other green toothbrush options: Life Without Plastic's toothbrush has a wooden handle with pig bristles and is fully compostable, and there are other toothbrushes made from bamboo (The Environmental Toothbrush, Green Panda), recycled plastic (Preserve), or those with replaceable heads (Terradent and Radius Source). But for my money, I'm sticking with a guy who I know is constantly working to make his product even more sustainable, who doesn't use animal products, and who is open to new ideas.

Plastic-Free Toothpaste, Powder, Soap, and Tablets You may not need toothpaste in a plastic tube for brushing your teeth. Here are a few interesting options:

- **LUSH Toothy Tabs.** They are aspirin-sized tablets that come in a cardboard box with just a couple of tiny plastic stickers to seal it shut. You crunch one tablet between your front teeth and start brushing. The tablet foams up just like regular toothpaste and cleans really well. Toothy Tabs are pricy, but I find I don't need to use a whole tablet at one time. Cutting them in half makes one box last twice as long.
- **Aquarian Bath Tooth Powder.** This tooth powder comes in a metal tin, or you can buy refills in glassine envelopes. The powder is black because it contains activated charcoal and bentonite clay, two effective cleaning agents, but it washes away completely and leaves your mouth feeling great.
- **Rex Apothecary tooth powder** comes in a 4-oz amber bottle, or you can order an 8-ounce refill in a recycled paper bag for the same price. It contains only 5 ingredients: bentonite clay, calcium carbonate, baking soda, peppermint essential oil, and stevia leaf for sweetness.
- **Make Your Own Tooth Powder or Paste.** The simplest homemade formulas are simply a combination of baking soda and salt. Add coconut oil or vegetable glycerin (if you can find it in a glass bottle instead of plastic) to make a paste. Add essential oils, Xylitol, or stevia for flavor. The Internet is full of DIY recipes. Google is awaiting your command.
- **Rose of Sharon Acres Tooth Chips.** Growing numbers of people swear by tooth soap. And Rose of Sharon Acres has added flavorings and sweetener to make their tooth chips taste less like soap and more like something you'd voluntarily put in your mouth. The chips come in a little metal tin. Personally, I couldn't get past the soapy taste, but then, I don't like cilantro either. You might love tooth chips!
- **Plain Soap** Hardier souls might skip the flavoring and extra expense of tooth chips and simply brush with plain soap. It's certainly one of the easiest ways of brushing your teeth. I gave it my most valiant effort and couldn't do it

without triggering my gag reflex. But many people brush with soap and prefer it to toothpaste. We're all different.

- **Plain Water.** Some dentists claim that you actually brush more thoroughly without any product on your brush at all and that what's important is the brushing and flossing. I'm not a dentist, and this suggestion should not be construed as medical advice.
- **Tom's of Maine Toothpaste.** If none of the choices above appeal to you and you plan to stick with regular toothpaste, why not choose Tom's of Maine? It comes in a plastic tube, but like Tom's of Maine deodorant, it can be downcycled through Terracycle's Nature Care brigade.

Dental Floss with Less Plastic No dental floss is completely plastic-free. Here are a few options.

- **Radius Natural Silk Dental Floss** is biodegradable but comes in a plastic container. Silk floss may not be appropriate for vegans.
- **Eco-Dent Floss** is made from nylon and comes on a plastic spool inside a cardboard box. Eco-Dent has the least plastic packaging, but the floss itself is made from plastic.
- **Floss with Thread?** Several people have told me they floss their teeth with cotton or linen thread instead of dental floss. I haven't tried it and am not sure I will. But it is an idea.

Whatever dental floss you choose, make sure the ingredients are listed on the package and that you know what the floss or tape is coated with. Many commercial dental tapes these days are made with perfluorchemicals to help them slide between your teeth.

Makeup

Beginning in seventh grade, for the first two weeks of every school year, I would spend an hour in the morning blow drying and curling my hair to look like Farah Fawcett and applying gobs of makeup: foundation, cover up, eye shadow, eye liner, mascara, blush, powder, and lip gloss. I'd promise myself that this school year was going to be different. I was going to fix my hair every day. I was going to wear makeup everyday. I was going to be popular. And slowly, as the weeks passed, my desire to fit in would wane as my natural laziness took over. By the end of the year, I was lucky to throw on matching socks before running out the door with my hair in a scrunchy and my face naked as when I was born.

I wish I could have loved my unadorned face back then. I wish I could have recognized it as a sign of health that I wasn't poisoning myself with toxic chemicals to be one of the cool kids. But even today, as I stare in the bathroom mirror at the brown spots on my face from too many sunburns at the beach and the creases worn into my brow from years of squinting, I think, "Botox would fix that right up," even as I know I would never really let anyone inject botulism into my skull. Still, the pressure is there to look the age that I usually feel, which is closer to fifteen than fifty, even if I'm too lazy to follow through.

I thank the universe for my laziness because natural, plastic-free makeup is hard to find. Although there are a handful of cosmetics companies producing healthy, high quality makeup in plastic-free packaging, the products tend to be expensive compared to what you'd find at the corner drugstore. And really, they should be. Maybe instead of using makeup to disguise what we actually look like, we could choose just a few products and apply them sparingly and judiciously to enhance the natural beauty we already possess. After all, as my makeup-free friend Sui Solitaire says, "We're so much more than what our physical bodies look like." It's true. Still, it's sometimes fun to apply a little paint.

Choose Makeup Brands Packaged with Less Plastic. Two companies producing organic makeup in plastic-free packaging are RMS Beauty and Kjaer Weis. Both companies were founded by NY-based makeup artists who got tired of the toxic cosmetics they were dealing with on a daily basis and thought they could do better.

- **RMS Beauty** founded by Rose-Marie Swift, offers cream-based lip shines, lip and skin balms, lip2cheek color, eye shadows, 'un' cover-up, "living luminizer," and raw coconut cream, all made from unrefined, raw, organic ingredients packaged in little glass pots.

- **Kjaer Weis** founded by Kirsten Kjaer Weis, offers a refillable makeup system, which includes beautiful reusable metal compacts that can be refilled with makeup pans packaged in recyclable cardboard. Products include cream blush, lip tint, and eye shadow.

Look for Handmade Cosmetics from Etsy Sellers. and request the seller package the products in plastic-free containers and ship without plastic packaging.

- **T.W.I.N.K. Beauty On Etsy.com** offers the only plastic-free mascara that I have found. It's an old-fashioned "cake mascara" that comes in a metal tin and is applied with a lash brush. If you already have a lash brush from an old tube of mascara, you can request "no lash brush" when placing your order and avoid that bit of extra plastic. Cake mascara takes a little getting used to, but it works great. T.W.I.N.K. Beauty also offers metal tins of eye shadow, blush, lip color, and finishing powder.

Making homemade face powder out of kitchen ingredients

Make Your Own Cosmetics. Here are a few homemade concoctions I devised in my kitchen. Also, check out the list of DIY makeup recipes on the Campaign for Safe Cosmetics website (Google "Campaign for Safe Cosmetics DIY Recipes") or peruse the natural makeup recipes on the Crunchy Betty blog.

- **Loose Face Powder.** Add to a base of cornstarch or arrowroot powder enough cocoa powder and/or cinnamon and/or bentonite clay powder to reach the desired color. Bentonite clay is green, so it will reduce red tones in the face. A mortar and pestle helps, but it's not essential. I made up a batch of this powder before a big presentation that I knew would be recorded because I didn't want my face to look shiny on camera. It worked.

- **Powder Blush.** Add to the above recipe just enough beet juice to reach the desired color. A little goes a long way. Cook beets in the microwave until they release their juice. Beet juice stains, so be careful not to get it on your clothes! Use within a week because the beet juice will turn brown over time.
- **Lip Gloss.** Plain coconut oil makes a great lip shine. Add some beet juice for a little color.
- **Eye Makeup Remover.** Plain coconut oil. Apply with a reusable cotton round.

Medicines & Supplements

One kind of plastic I haven't seemed to be able to avoid completely is prescription bottles, which cannot be refilled in the state of California, so they end up in the recycle bin. Gimme5 will take back any bottles with a #5 on the bottom, but that doesn't address the caps or bottles made from different kinds of plastic. As for over-the-counter supplements and remedies, many of them can be homemade using fresh ingredients and dried herbs. My local Whole Foods Market and other natural food stores carry a large selection of bulk herbs that can be purchased using our own containers. Books like *The Herbal Home Remedy Book*, by Joyce A. Wardell (Storey Publishing), and websites like Prevention's Home Remedy Finder contain all kinds of ways for dealing with everyday ailments, many of which include strategies and ingredients that don't require plastic packaging.

Feminine Hygiene & Adult Personal Products

Instead of disposable pads, tampons, and menstrual cups made with and packaged in plastic, consider switching to products that are either compostable or reusable.

Make Your Own Cough Syrup.

Mix together and store in a small jar:

- ¼ teaspoon (1 g/1.23 mL) cayenne pepper (bulk bin)
- ¼ teaspoon (1 g/1.23 mL) ground ginger (bulk bin)
- 1 tablespoon (15 mL) honey (bulk or glass jar)
- 1 tablespoon (15 mL) apple cider vinegar (glass bottle)
- 2 tablespoons (30 mL) water (tap)

Another option recommended by my friend Tracey TieF is a combination of ginger, lemon, and slippery elm, which you might be able to find with the bulk herbs.

Compostable Pads, Liners, and Tampons Natracare offers a full line of organic cotton compostable pads, liners, and tampons that are manufactured without dioxins or petroleum-based plastics.

Reusable Cloth Pads and Liners Reusables are an even more sustainable solution, since they can be washed and used over and over again. Nowadays, there are many different companies offering cloth pads and liners in different patterns, shapes, fabrics, and thicknesses. Whether you choose 100 percent plastic-free pads or pads with some waterproof material, using cloth pads cuts down on a huge amount of plastic waste compared to disposable pads. Here are just a few options:

- **Lunapads** offers a full selection of 100 percent cotton, plastic-free pads and liners (including thong-shaped), which are available in regular or organic cotton. The company also offers optional Lunapanties, which are specially designed to hold pads in place. Check out the website for handy How-To videos that explain how to use, wash, and care for cloth pads.
- **GladRags** offers an assortment of 100 percent cotton, plastic-free pads and liners, which are available in various patterns, including undyed organic cotton.

- **Lollidoo** is a cloth diaper company that also offers three sizes of reusable cloth menstrual pads made from organic cotton and recycled fleece. The fleece is made from recycled #1 PET plastic, like soda and water bottles.
- **Cloth Pads on Etsy.com** are available from many different sellers in a vast array of sizes, patterns, fabrics, and styles. Different sellers use different amounts of waterproof materials (read: plastic), so check the descriptions carefully.
- **DIY Cloth Pads** could be easy and fun, if you're crafty. Search Google for "make your own cloth pads" for a plethora of patterns, instructions, and tips.

Reusable Menstrual Cups are nonabsorbent devices inserted into the vagina to collect menstrual blood. Unlike cloth pads, they don't require you to do extra laundry, and they can be worn while swimming and engaging in all those other activities that disposable tampons supposedly allow. They can be worn for up to twelve hours (depending on your flow), after which you simply empty the cup, wash it with mild soap and warm water, and re-insert. Here are a few choices:

- **DivaCup** and **Moon Cup** are made from medical-grade silicone, a material I'll discuss more in chapter 9.
- **The Keeper** is made from natural gum rubber (latex). The Keeper website offers both The Keeper and the Moon Cup in plastic-free packaging.

Note: Do not confuse these reusable options with devices like Instead Softcup, which is made from disposable plastic and is meant to be thrown in the trash after a single use.

Bulk herbs from Rainbow Grocery

Personal Lubricant Coconut oil makes a fantastic personal lube and is said to contain antifungal properties that are helpful for women who are susceptible to yeast infections. Other people swear by olive oil. But keep in mind that oil-based lubes do not play well with latex. If you use latex condoms or a diaphragm, you'll want to consider water-based options. My friend Tracey TieF suggests mixing water with slippery elm powder, xanthan gum, or guar gum, any of which you might be able to find in bulk.

Diapers

According to the EPA, Americans discarded nearly 3.6 million tons of disposable diapers in 2012.[103] That's a whole lot of wasted paper and plastic, but just like feminine hygiene products, disposable diapers can be replaced with reusable cloth options.

Cloth Diapers Since I don't have kids, I called on my friend Calley Pate, who blogs at The Eco Chic and works for DiaperShops.com, to explain to me the ins and outs of cloth diapering. During a fun Skype conversation between our homes in California and Florida, Calley used a Cabbage Patch doll to demonstrate for me the various types of cloth diapers available these days (we've come a long way since the plain rectangular prefold diapers our moms might have used when we were babies) and how they work. Today's diapers are fitted, come in adorable patterns and styles, and generally use snaps or Velcro instead of pins. And gone are the days of those big plastic pants you had to put on over the cloth diapers to keep them from leaking. The conversation lasted several hours, after which I felt like I could write an entire book on the subject. Fortunately, I don't have to because the definitive cloth diapering book was recently published in 2011: *Changing Diapers: The Hip Mom's Guide to Modern Cloth Diapering*, by Kelly Wels (Green Team Enterprises). If you're curious about cloth diapering, check out the book. Here are a few helpful online resources:

- **The Real Diaper Association** is a cloth diapering advocacy group that provides resources and information about cloth diapering. Read about its Cloth Diapering in Daycare Campaign and the environmental impacts of cloth vs. disposable diapers.

- **The Diaper Pin** is a message board for discussing any and all topics related to cloth diapering. Learn from the experts—other parents. Among other tools, the site includes a Cloth Diaper Savings calculator, so you can figure out how much money you would save by switching to cloth.
- **Diaper Swappers** provides a marketplace where people can save money and reduce their cloth diaper impact further by buying, selling, and trading secondhand cloth diapers. The site also provides a discussion forum for parents to post their cloth diapering questions and to discuss many other parenting issues.
- **The Diaper Jungle** provides resources for learning to sew your own cloth diapers, as well as a directory of cloth diaper stores and a discussion forum.
- **Dirty Diaper Laundry** provides video reviews of various brands of diapers so you can see what they look like and how they work before purchasing.
- **Cloth Diapering 101** Check out the series of posts on the Eco-Novice blog labeled "Cloth Diapering 101," wherein Betsy addresses such topics as why she loves cloth diapers, her favorite daytime and night-time diapers, what to do with the *poop*, using cloth wipes, and how to know if cloth diapers are right for you.

Cloth diapers run the gamut, from those that are made of 100 percent organic natural materials to those made of synthetic fabrics. Here are a few resources and ideas for the least plastic cloth diaper and accessory options:

- **Firefly Diapers** sells organic cotton diapers and training pants; organic wool diaper covers, changing mats, mattress pads, and puddle pads.
- **sustainablebabyish | sloomb** sells organic cotton/bamboo diapers and wool covers.
- **Lollidoo** sells diapers, covers, and bags made from organic cotton and fleece made from recycled beverage bottles.
- **Kanga Care** offers an "Eco-Posh" line of cloth diapers made from organic cotton, bamboo, and recycled bottles.
- **Cotton Babies** sells many different kinds of diapers. Click the "Organic Diapers" link to find diapers and accessories made from organic cotton and hemp.

- **Recycled wool "butt sweaters"** are water resistant diaper covers made from felted, repurposed sweaters. A Google search on "butt sweaters" brings up a huge list of instructions for making your own. Or search for "recycled wool soaker" or "recycled wool diaper cover" on Etsy.com for an assortment of choices. Why consume new resources when we can reuse what already exists?

A study from the U.K. Environment Agency comparing disposable diapers and cloth diapers for environmental impact concluded that cloth diapers' biggest environmental impact comes from washing and that we can reduce the impact of cloth diapers by washing full loads of laundry, line drying when possible, using energy efficient appliances, and opting for secondhand diapers instead of buying them new.[104]

gDiapers An option between cloth diapering and disposables is the hybrid gDiaper system, which uses a disposable, nearly plastic-free diaper refill (gRefill) inside a reusable cloth pant (gPant). The gRefill is made from a cellulose rayon coverstock; a fluffed wood pulp interior; and a small amount of super absorbent sodium polyacrylate (SAP) granules. The diapers themselves come packaged in plastic, and the reusable cloth gPant contains a polyurethane-coated nylon liner, but overall, the system generates much less plastic waste than conventional disposable diapers.

Elimination Communication One way to get babies out of diapers quicker and reduce the need for either disposables or cloth is to learn to recognize when they need to eliminate and help them communicate the need in advance. "Elimination communication" techniques can begin in early infancy so that children are "toilet independent" before most children are just beginning conventional potty training. Taina Uitto, author of the blog "Plastic Manners," has used this method successfully with her baby. She wrote me:

> I started at three months old. I put my boy over the toilet and made a pee sound. I did this as soon as he woke up, and about 5–10 minutes after eating... repeat. You will learn their schedule quick; much quicker if you don't use diapers in between. At age six months I know his routine well, and he knows what to do at the potty. If he has to

pee, he goes right away. He goes number two first thing in the a.m., so no need to worry for the rest of the day. (Plus it's so easy to clean—no need for wipes.) He is very proud when he goes. He also knows to tell me if he really doesn't have to go (he looks up at me). If he starts to pee elsewhere, he can stop in the middle of it to get to the potty. Just make a surprised sound and they naturally stop.

I think most women don't even consider diaper-free because we are taught babies wear diapers, sometimes for many years (!!) (Big $$ in diapers and wipes). Babies are much more comfortable without, and you will also save money (and plastic) on creams, not to mention your baby from excess products.

Visit www.diaperfreebaby.org for more information.

Toilet Paper and Other Disposable Paper/Cotton Products

Reducing our use of disposable paper products is a good idea in and of itself, as paper has a huge environmental impact. But when I first began my plastic-free project, I was more concerned about the plastic bags and wrappers that those products are packaged in. I searched high and low for toilet paper, facial tissues, napkins, paper towels, paper plates, as well as disposable cotton products like swabs, rounds, and balls in plastic-free packaging. Thankfully, I found a nearly plastic-free solution for toilet paper, as I'm not willing to give that up (although there are alternative methods), and as for the rest of it, I've replaced it all with reusables so I can cut out the disposable paper altogether.

Bathroom Tissue and Other Plastic-Free Ways to Wipe

- **Seventh Generation** offers its toilet paper in either plastic packaging or individual rolls wrapped in thin tissue paper. Because buying individual rolls from the store can be costly and cumbersome, I order them a case at a time from Amazon.com. Each 60-roll case comes in Amazon's "frustration-free packaging," which means a plain cardboard box with a little bit of plastic shipping tape, as opposed to the big plastic overwrap in which most paper products come encased. What's more, I save an extra 15 percent by subscribing through Amazon's "subscribe and save" program, which means that a new case is shipped to me every few months, unless I opt out of a particular delivery. If you want the extra discount, be sure to order the cases sold from and shipped by Amazon.com and not a third party seller.

- **Bumboosa** offers toilet paper made from 100 percent bamboo, instead of trees or recycled paper, and packaged in a cardboard box. Bamboo is a fast-growing grass that regenerates much more rapidly than trees. Bumboosa toilet paper is manufactured using a thermo-mechanical pulping process that has a lower impact than the chemical process used to produce most toilet paper. The main drawback to this product is that it is much more expensive than other brands of toilet paper, but hopefully demand for tree-free bathroom tissue made from grasses will increase and drive the price down.
- **Family Cloth** are reusable cloth wipes used after toileting instead of disposable toilet paper. Some women use it only after peeing, while other families use it to wipe after Number Two as well. Moms who are accustomed to cloth diapering say it's the same principle, and really no big deal. Some people make their own cloth wipes from old t-shirts and other scraps of fabric or order them via Etsy sellers. Google "family cloth" for lots of articles and blog posts about using reusable cloth wipes after toileting.
- **Water** is used throughout the world instead of toilet paper or wipes for cleaning fronts and backs, and many people are installing bidet attachments onto their toilets to avoid toilet paper waste. A device like Brondell's Swash EcoSeat is powered solely by water pressure, rather than electricity, and while the contraption itself is made from plastic, it can potentially save a lot of waste in the long run. Of course, in India and many places in North Africa and the Middle East, cleaning after toileting is accomplished with simply the left

BPA found in recycled toilet paper. While it's important to me to save trees by using postconsumer recycled toilet paper, I am concerned about the trace amounts of BPA that have been detected in recycled toilet paper[105] and other recycled paper products, probably as a result of BPA-coated thermal paper that consumers mistakenly toss in with the regular paper recycling. (As I mentioned in chapter 4, thermal paper should go in the trash, not the recycling.) To me, this is one of those areas where we as consumers have no clear environmental/health choice. We need legislation prohibiting the use of BPA, so contamination of recycled paper will no longer be an issue. In the meantime, I go ahead and use recycled toilet paper, reasoning that I'm cutting out BPA in other aspects of my life. You may feel differently. Fortunately, there are a few other alternatives.

hand (which is sanitized with soap and water afterwards) and a pitcher of water and is considered to be much more sanitary than wiping with plain paper. While I'm not willing to go there, I have a friend who swears by this method.

Handkerchiefs Instead of Kleenex I switched from disposable facial tissues to cloth handkerchiefs to avoid the little plastic window in the Kleenex box but came to realize that saving all that paper was even more important. I've found some sweet embroidered hankies at local thrift stores, and friends of mine hand make them from old t-shirts or flannel sheets. Etsy sellers offer a plethora of choices. The HankyBook is an ingenious system of organic cotton "pages" bound together like a little booklet, which keeps your handkerchiefs neat and tidy. Learn to use a neti pot to keep sinuses cleaned out on a regular basis. I love mine, which I finally learned how to use after watching several demonstration videos on YouTube. Seems people love to record themselves doing strange things like pouring water through their noses.

Reusable Cloth Facial Rounds Instead of buying plastic bags of disposable cotton balls, why not invest in some reusable cotton rounds? They are great for removing makeup, dabbing on antiseptics, applying deodorant powder, and almost anything else for which you would normally use a disposable cotton ball or cotton round. (Don't use them to wipe off nail polish, as the polish will not wash off.) Search for "organic facial rounds" on Etsy to find a variety of facial rounds, from soft organic cotton fleece to exfoliating hand crocheted versions. One of my favorite Etsy sellers is Tiffany Norton at Juniperseed Mercantile who is committed to packaging all her products plastic-free.

Natural Cleaning Cloths and Scrubbers Instead of paper towels, plastic scrubbers, synthetic sponges, or microfiber cloths, I've found some great natural, reusable alternatives.

- **Skoy Cloths** are one of my favorite plastic-free finds. Made from cotton and tree cellulose, each reusable cloth can take the place of 15 rolls of paper towels. They are super absorbent (holding up to 15 times their weight in water) and dry out very quickly if you squeeze and hang them after each use, eliminating the wet sponge ick factor. Skoy cloths can be cleaned in the washing machine or dishwasher or even sanitized in the microwave. And at the end of their lives, they can be composted. I'll bet you could just bury them in the backyard, but I actually wouldn't know because we've been using the same cloths for years, and they're holding up just fine. And Skoy cloths are not wrapped in plastic packaging, like so many other sponges and scrubbers are these days.
- **Compressed Natural Cellulose Sponges** are often sold in natural grocery stores without plastic packaging because they don't need to be kept moist; they expand when wet.
- **Coconut Coir Scrubbers and Brushes** are a good alternative to plastic. You can often find them at natural grocery stores or through online retailers like GetnGreen.com, which carries a coconut coir vegetable brush, bottle brush, and even commode brush.

> "Buy a good pair of pinking shears to cut old clothes into replacements for paper towels, diaper wipes, and other single use products." —Jeanne Bruner, Blanchard, Idaho

- **Loofah Scrubbers** are made from the fibers of an edible gourd and are completely biodegradable. In fact, from my experience, they may start to biodegrade in your kitchen sink if you don't let them dry out. These scrubbers come in lots of fun shapes and patterns and can be used not only to scrub pots and pans but to exfoliate your skin as well.
- **Copper Scrubbers** are great for cleaning severely baked-on messes. Copper, of course, is a mined metal with a higher environmental footprint than plant-based scrubbers, so use it sparingly. Chore Boy brand copper scrubbers come in a plastic-free cardboard box.
- **Handmade Rags and Cloths** can be used for everything from wiping up messes to polishing furniture.
- **Reusable Swiffer Pads.** Are you the proud owner of a Swiffer mop system? Are you tired of spending money and wasting resources on disposable mop pads that come in plastic packaging? So are quite a few Etsy sellers. Search for "Swiffer" on Etsy.com for an assortment of reusable, washable pads handmade to fit your Swiffer mop. My favorite comes from Juniperseed Mercantile.

Cleaning Products

When it comes to cleaning, less is more. Not less cleaning but fewer products. Scanning the shelves of the cleaning products aisles, I am amazed by the huge assortment of formulas, each in its brightly colored plastic bottle, meant for specific purposes. In my quest to live plastic-free, I've learned that we don't need so many different chemicals and that just a few natural products can do a lot of different jobs. What's more, we certainly can live without all the strong artificial fragrances those products contain, many of which contain phthalates to help them linger. You know what clean smells like? Nothing. Clean smells like nothing.

The Internet is chock full of recipes and suggestions for natural cleaners. And Rachelle Strauss's book *Household Cleaning Self-Sufficiency* (Skyhorse) is a great reference. Or check out the green cleaning recipes at Women's Voices for the Earth. Here are just a few of my personal plastic-free, natural cleaning solutions:

Clean with Vinegar and Water. I use a mixture of 1 part white vinegar (which comes in a glass bottle) to 2 parts water with a few drops of tea tree oil as an all-purpose spray cleaner. I use it on countertops, windows, mirrors, appliances, and even to clean up after my rascally kitties when they forget that pooping *near* the litter box is not the same as pooping *in* it. I store it in a repurposed spray bottle. So far, I have not found a plastic-free spray bottle, and I'm fine with reusing what I already have. If you don't like the smell of vinegar, consider substituting vodka or grain alcohol instead.

Baking Soda is a Fantastic Scouring Powder and Deodorizer. In fact, I use it for regular hand dishwashing instead of liquid dish soaps that come in plastic bottles. To clean really baked on messes, I first let dishes soak overnight and then scour with baking soda and a copper scrubber. Baking soda is also great for removing coffee and tea stains from mugs, cleaning out the refrigerator, scrubbing the bathtub and toilet, freshening the litter box, and absorbing odors anywhere in the house.

Powdered Natural Dishwasher Detergents. They come in recyclable cardboard boxes. We alternate between Ecover and Seventh Generation in our home. And we use white vinegar as a rinse aid instead of buying commercial products. Check the Internet or *Household Cleaning Self-Sufficiency* for DIY recipes using baking soda, washing soda, and/or borax in the dishwasher.

Use Soapnuts to Clean Clothes, Dishes, Hands, and a Lot More. Soapnuts are the dried fruit of a tree called *Sapindus mukorossi* (Chinese Soapberry) and contain saponin, a natural surfactant that foams just like soap. Soapnuts clean without the need for additional chemicals. To use them in warm water laundry loads, you simply toss a few soapnuts in a cloth bag in the washing machine and let them do their thing.

To use them in cold water or for other liquid soap uses, boil water and steep them for several hours to make a liquid soapnut soak. Soapnuts smell funny, so some people like to add a few drops of their favorite essential oil, but it's unnecessary. Clothes washed with soapnuts come out of the laundry smelling just like . . . clothes. I like Eco Nuts brand soapnuts because the company packages them in cardboard boxes instead of plastic.

Use Powdered Laundry Products without a Plastic Scoop. As with dishwasher detergents, powdered laundry products come in recyclable cardboard boxes instead of plastic bottles. However, most of them include a plastic scoop. The scoop might be nice to have the first time, but why would you need a new one each time you buy laundry powder? Meliora K. is a new brand of laundry powder made from only three ingredients: sodium bicarbonate, sodium carbonate, and plain soap. Only the regular 64-oz size currently comes with a plastic scoop—all other sizes do not. And the owner let me know that she is currently looking for an alternative to the plastic scoop, but in the meantime, customers may ask for the scoop to be omitted from their orders. Ecover brand laundry powder pours without any scoop. And Aquarian Bath lemon laundry powder comes with a cute wooden scoop, but customers have the option of requesting no scoop at all.

Treat Laundry Stains with Borax or Solid Laundry Stain Bars. A paste of borax and water is a good stain pre-treatment, and Buncha Farmers makes an all natural stain-remover bar packaged plastic-free. There are many different Do-It-Yourself stain treatments for specific kinds of stains. Google can help you out in an emergency.

Remove Wine Stains from Carpet with Salt and Carbonated Water. What happens when you try to use your soda maker to carbonate red wine? You end up with a red wine volcano spewing all over your dining room table and carpet, that's what happens. But it's a good thing you have a soda maker to make carbonated water to clean up the mess. Here's how you do it: absorb most of the liquid with a Skoy cloth, old towel, or rag, but don't scrub yet; sprinkle salt on the stain right away to absorb as much remaining liquid as possible; when dry, vacuum up the salt; pour carbonated water on remaining stain and scrub with a brush. With a little elbow grease, the stain will come right out. I should know. I've had to do it multiple times.

Natural Rubber Gloves. are great for anyone who is not allergic to latex. Casabella makes a fantastic all purpose natural rubber glove lined with 100 percent cotton flocking and packaged in a cardboard box. The gloves have special cuffs that catch water and prevent wet sleeves and arms. The only drawback is that they are bright pink. I'm not sure if that's an indication that they are meant for women, but if you'd rather wear a green-colored glove, check out the natural rubber gloves from If You Care, which are made from Forest Stewardship Council (FSC) certified latex (i.e. natural rubber sourced from a responsibly managed plantation) with a 100 percent cotton lining. They also come in a cardboard box.

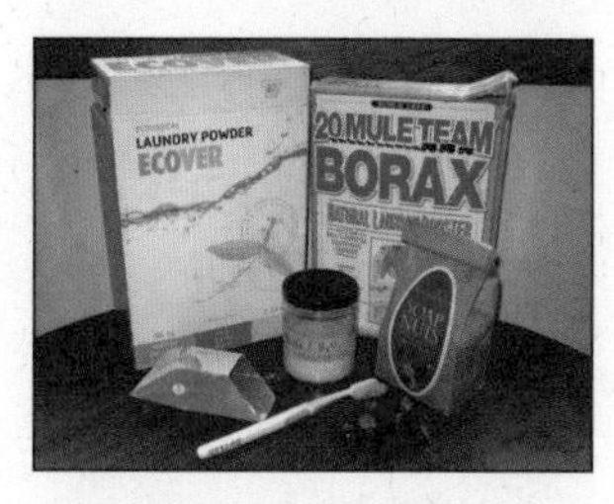

The Cat Litter Dilemma

When choosing a cat litter, it was important to us that both the litter and the packaging were eco-friendly and safe for our kittens. Clay litters wreak havoc on the environment and produce silica dust that is dangerous for cats and humans to breathe. Internet anecdotes suggest that clumping versions of clay litters may be fatal to small kittens when swallowed.[106] Manufacturers claim their clay products are safe, but I choose to follow the

Plastic-free cleaning products

Precautionary Principle. Why take the risk? Fortunately, there are many different kinds of litters made from biomaterials, including wheat, corn, wood pellets, and even recycled newspaper, but many of them come packaged in plastic bags. We ended up choosing SwheatScoop, which is a flushable clumping litter made from wheat that comes in a recyclable paper bag with no plastic liner. It's non-toxic and perfectly safe for cats if swallowed. We do have to keep the bag out of reach of Soots and Arya, who would be really happy if we just dumped it all over the floor and let them swim in it. We don't. We're mean like that.

A note about flushable cat litters: Environmental groups advise us not to flush cat litter or cat poop because it is likely to contain the parasite ***toxoplasma gondii***, which is deadly to sea otters, and municipal waste water treatment facilities are unable to remove the parasite before it enters our waterways. Cats pick up "toxo" from eating small wild animals. If you have indoor-only cats, you can have your vet test them for the parasite. If the test is negative ***and*** if the cat never goes outside or comes into contact with wild animals ***and*** if you never feed the cat raw meat, it might be safe to flush its poop and/or certified-flushable litter.[107]

Trash Bags—Dealing with Garbage and Pet Waste

The two most frequent questions I get from people trying to get the plastic out of their lives is what I use for garbage bags and what kind of bag to use for pet waste. As it happens, those are the two main reuses for plastic grocery bags. If stores stop giving them out, will people have to start buying trash bags and poop bags? I say no. We have other options, none of which include buying anything new.

Naked Trash Cans. Think about the waste paper baskets and other trash containers inside your home. How many of them actually need a liner in the first place? A liner might be useful for wet garbage, but if your waste cans mostly contain dry trash, do they actually need any kind of bag? Allocate one or two waste cans for wet garbage and keep the rest bag-free.

Reuse What You Already Have. If you're just beginning to get the plastic out of your life, you probably have many other kinds of non-recyclable bags and containers that could be used to pick up pet waste or line small trash cans. Bread bags, chip bags, cereal bags, bagel bags, tortilla bags, frozen food bags, and newspaper wrappers can all be used to contain pet waste or wet garbage. Cardboard milk cartons can be used for wet stuff as well.

Reuse Other People's Bags. You may have stopped accepting plastic grocery bags and produce bags, but that doesn't mean your friends and neighbors have. Consider putting an ad on Freecycle.org or asking around for other people's used bags and wrappers. And while you're at it, mention to them that it would be okay if they stopped taking bags from the grocery store and started carrying their own. In a nice way, of course.

Use Old Newspaper. Once you start getting away from the plastic habit, you'll probably want to transition away from plastic trash bags entirely. Sheets of old newspaper work well for picking up pet waste and for small amounts of wet garbage. Of course, paper and other bio-based materials are better in the compost than the landfill, so consider the next step . . .

Start Composting. We don't use any garbage bags in our home at all anymore, and composting is the reason why. If you can compost all your food waste, there really isn't a need for trash can liners. We keep a compost pail on the kitchen counter and fill it with all of our food scraps. Then, we have a choice: we can either empty the pail into our city's green compost bin, from which it will be trucked to an industrial compost facility, or into our own home compost. Visit www.howtocompost.org to learn everything you need to know to get started composting. Think you need a yard or more room? There are options even for people who live in small spaces. Composting can

"One old phone book makes about 1,000 pooper scoopers. Of course, it helps to have a small dog with small poop, but you can use more pages and make them for bigger dogs. I fold the pages into a cone like a paper hat." —Autumn Dann, Berkeley, California

> "I live in a small studio apartment with no yard or room for a composter. Still, I compost. I save up all my compostable scraps (eggshells, coffee grounds, vegetable peelings, dead leaves from houseplants, etc.) and put them in a bag I keep in my freezer. A non-recyclable chip bag is good for this. Then, every couple of weeks, I take my scraps to a community garden that has several composters going. I bring the bag home and wash it out and start collecting food scraps again."
> —Cat Domiano, New York, New York

help save plastic garbage bags, but even more important, it keeps methane-producing food out of the landfill.

- **Traditional Compost Bin.** These bins are for people with yards because the compost in the bin comes in contact with the ground. You can buy a home composter or make your own. A compost bin can be as simple as a trash can with the bottom removed.
- **Compost Tumbler.** A tumbler is suspended off the ground. To turn the compost, you simply rotate the tumbler. Some compost tumblers work well on decks and patios in places without a yard.
- **Worm Bin.** Even apartment dwellers can maintain a worm bin, which requires much less space than a regular composter and relies on little critters to do most of the hard work of turning food scraps into compost. Check out www.redwormcomposting.com for everything you ever wanted to know about vermicomposting.
- **Nature Mill.** This is an electronic composting system that is small enough to fit in a kitchen cabinet and creates the compost for you. Nature Mill is expensive but seems like it could be a viable option for people without access to municipal composting facilities who have no space for a bin and no time to maintain one.
- **Compost Heap.** This is the lazy person's compost system and one that I've regressed to

recently. For a few years, Michael and I maintained our own home compost bin until we, and by "we" I mean "I," got too lazy to keep it going and started using the city compost bin instead. Then, after starting my garden, I decided I wanted to keep my food scraps to feed my own plants. But, still too lazy to start up a real compost system again, I designated a spot in a corner of my tiny yard to be the compost heap. I empty my produce scraps, coffee grounds, and ground up egg shells onto the pile every day or so and cover it up with straw mulch that I bought for my garden. It's a slow process, but it works.

- **Community Compost Facility.** If you don't want to make your own compost, find out if there is a commercial facility or community garden that will take your food scraps. Visit www.findacomposter.com for a directory of composting facilities throughout North America.

"I collect bagged leaves from my neighbors' curbs in the fall, use the leaves in my compost pile, dry out the plastic bags, and save them for re-use. Once I have enough, I make packets of folded bags to return to the homeowners. This process saves leaves and plastic from the landfill."
—Betsy Robertson, Master Composter, San Marcos, Texas

Compost Pet Waste. Instead of throwing dog poop in the garbage, consider composting it instead. Pet waste is not allowed in most municipal compost systems, but you can compost it yourself in a part of the yard that is far away from edible plants or running water. Purchase a Doggie Dooley in-ground pet waste toilet or Google "how to make a dog poop composter" to learn how to build your own. Another option is to scoop up dog poop in a bucket and bring it home to flush.

Kitchen compost pail

A note about biodegradable garbage or pet waste bags: If the garbage or pet waste is destined for the landfill, don't waste your money on biodegradable bags. As I've mentioned before, biodegradability is not desirable in a modern landfill anyway. Certified compostable bags can be helpful for containing food scraps and pet waste that will be composted. However, if your compost goes to a commercial facility, make sure compostable bags are allowed. Many facilities that produce certified organic compost do not allow bioplastics at this time.

QUICK LIST: QUESTIONS TO ASK

Here's a quick list of questions to ask yourself when deciding what personal care and household cleaning products to buy.

1. Can I find the product in solid or powdered form without any plastic packaging? Have I checked resources like Etsy.com?
2. If not, can I make it myself from packaging-free ingredients?
3. If not, can I find the product in bulk so I can fill my own container?
4. If not, can I substitute something else? Maybe there is an alternative which is similar enough or would satisfy the same need.
5. If not, do I really need it in the first place?
6. If yes, can I find it in compostable cardboard, metal, or glass?
7. If not, can I buy it in a large size with less plastic packaging?
8. If not, can I find it in packaging that is easily recyclable where I live?

Personal Worksheet: Personal Care/Cleaning Substitutions

Use this sheet to list the top 20 personal care and/or cleaning products you buy regularly that come packaged in plastic. As you discover a plastic-free solution, mark it on the list. Solutions could include a different brand, store, recipe, alternate product, or even doing without. Use the list to make one change every two–four weeks, depending on availability of plastic-free products in your area.

Type of Product	Brand I Buy Now	Less-Plastic Alternative

Going Further: Asking Companies to Change Their Packaging

It may seem like writing or calling a company to request plastic-free packaging won't do much good. Often, a customer service representative will take your message or send you an email response that is little more than a standard form letter. But all of our calls, letters, and emails add up and make an impact. I already shared with you the success that I and 16,000 other people had letting Clorox know we wanted to be able to recycle our Brita filters. But I've had successes on smaller levels too. One reason I love to support small businesses and Etsy sellers is that you can usually contact the owner directly before making a purchase and find out if a product can be packaged and shipped plastic-free. The small business relationships I value the most are with owners who listen to my suggestions and change the way they package their products for all of their customers, not just for me. One such business owner is Lisa Albrecht from LaundryTree. Unfortunately, LaundryTree went out of business in 2014, but this remains one of my favorite success stories.

A few years ago, I read about LaundryTree soapnuts on the green living blog Tiny Choices and noticed in the photo that the product seemed to be packaged in plastic bags. I left a comment on the blog asking if that were the case; the blog writer contacted LaundryTree; and Lisa from LaundryTree contacted me. She said that she actually wanted to move away from plastic bags but wasn't sure what packaging to use instead, and she asked if I would help her figure it out. Wow. I was stunned! I honestly hadn't expected such a positive response. And I'm not sure how much help I actually provided Lisa besides giving her the encouragement and support she needed to do the research. In less than a month, she had found her solution and started packaging her soapnuts in recycled kraft paper bags lined inside with PLA bioplastic instead of fossil-based plastic. As she phased out the plastic bags she already had, she switched over completely to the new bags. A huge company like Procter and Gamble or Clorox would have required months of product design, focus groups, and retooling before making a change in packaging. Small companies like LaundryTree have the flexibility to make changes quickly.

Because of successes with companies like LaundryTree, I continue writing to companies and asking for change. I've even been known to mail back unexpected plastic laundry powder scoops. I don't always know if my letters or emails have made an impact, but what I do know is that most companies are not going to change their policies and practices unless we speak up. So let's speak up.

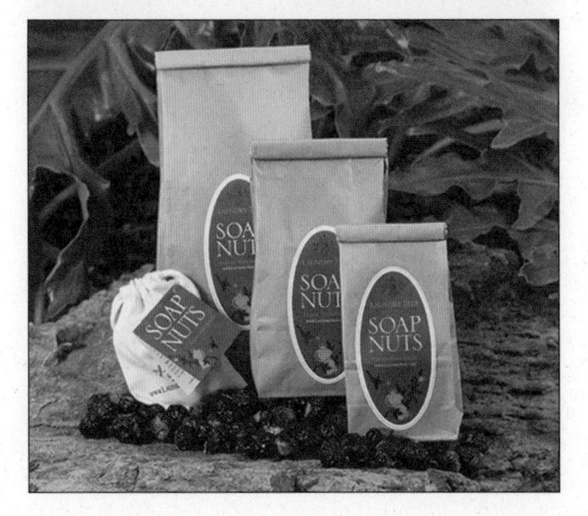

Action Items Checklist

(Choose the steps that feel right to you. Then, as an experiment, challenge yourself to do one thing that feels a little more difficult. Only you know what that one thing is.)

- ☐ Look up the cosmetics you currently use on the EWG Skin Deep Cosmetics Database to decide which ones, if any, to phase out.
- ☐ Visit BeatTheMicrobead.org to learn what products contain microbeads and eliminate them from your home.
- ☐ Locate stores selling personal care/cleaning products in bulk.
- ☐ Use Personal Care/Cleaning Substitutions Worksheet to make a plan to eliminate one plastic-packaged item every 2–4 weeks.
- ☐ Talk to a store manager about adding more bulk/plastic-free personal care/cleaning items.
- ☐ Write to a company about switching to plastic-free packaging.
- ☐ Sign up with the Campaign for Safe Cosmetics or Safer Chemicals, Healthy Families to work for updated toxic chemical legislation.

Chapter 8: Feeling Overwhelmed (Help! I'm Drinking My Cleaning Fluid!)

I know what you might be thinking at this point. "Plastic is everywhere! It's all so overwhelming!" That's true. It is. And I haven't even gotten to computers and cell phones and toys and clothing and appliances and carpets yet. So let's take a break and talk about feeling overwhelmed.

I mentioned in the previous chapter that vodka—or grain alcohol like Everclear—makes a great cleaning fluid. It worked great for me. That is, until I started drinking it every night. I drank it because it was there and provided an easy escape when I felt like planetary problems were beyond my control. Somehow I had allowed myself to become overwhelmed with plastic and blogging and saving the world. Every day I left my house and saw plastic bags blowing down the street, plastic bottles rolling across the parking lot, plastic caps and wrappers and straws teetering on storm grates ready to drop into our waterways. And I felt helpless.

Shopping in stores like Whole Foods was no longer the fun fact-finding mission it once was. Instead, rows of plastic-packaged cheese, chips, candy, cookies, crackers (and I'm still only in the Cs) taunted, "You'll never get rid of us." And the irony of customers filling up their reusable bags with plastic-packaged processed foods sent me over the edge. *Blame* and its cousin *Guilt* were beating the crap out of me. I couldn't allow myself to sit still for a minute because my own mind had become a bigger enemy than the plastics industry. And although I had given up some of the crutches I previously used to cope (fast food, retail therapy), there were still plenty of others to lean on.

One night, around 2 AM, I found myself at the twenty-four-hour Long's Drugstore down the street, frenetically pacing up and down the aisles while sipping furtively from a stainless steel Klean Kanteen full of red wine—wine from a glass bottle with a natural cork stopper, of course—and desperate to find something not made out of plastic that I could buy. I knew I was out of control, and yet I couldn't stop. Up and down the rows of cleaners and bathroom fixtures and cookies and photo albums and shampoos and water pistols and antacids and laundry baskets I rolled my cart. I ended up purchasing a wool duster that I knew I'd probably never use. Why? Because I just wanted to ***buy*** something. Anything. Like so many other people in our crazy fast-paced society, I wanted to fill up that sense of overwhelm with ***stuff***.

I don't remember what happened when I got home that night. I'd like to be able to say that I sat down and had a good cry and the kind of spiritual breakthrough to which journeys like mine are supposed to lead. But I'm guessing I probably collapsed into bed and woke up the next morning feeling crappy and just as miserable as I had been the night before. I do remember that I called my friend Mark and told him the story, and we both had a rollicking good laugh at the thought of this crazy drunk eco-woman weaving through the aisles of Long's Drugs in the middle of the night, chugging from a stainless steel bottle. But the laughter was insincere. I wanted someone to fix me.

See, I knew there had to be a way—a way for me, for all of us, to see clearly the immensity of the challenges we face without giving up, still doing our work. And eventually I turned to No Impact Man Colin Beavan for advice. During one of his visits to San Francisco, I asked him how he managed to cope when the problems in the world just seemed so massive. And he said, "There is such a thing as being overwhelmed and feeling the feelings, and also putting one foot in front of the other. ***At the same time.***" And then he told me this story:

> There's a woman who's a layperson, a Zen adept, as well as a grandmother. All her life she's been a lay preacher. And then her granddaughter dies. And she's sobbing her eyes out. And all the people are so surprised, and they say to her, "Don't you understand that everything is one? Everything is just the way it's always been. Do you not understand the point?"
>
> And she stops crying and says to them, "Don't you understand that my tears save my granddaughter and all beings?" That is to say my compassion, my sadness for the human condition is what makes me human. So, that in itself is a great story, right?

> But the really important part of that story is that she stops crying. In other words, she's overwhelmed with sadness, but when somebody comes to her who needs to be taught, she puts one foot in front of the other. She stops crying and she teaches. We have to accept that the problem is overwhelming and immense and at the same time just get on with it.

I've developed a few strategies for when I feel overwhelmed by the immensity of the environmental problems we face: things to do instead of reaching for a bottle of booze or sleeping pills. Some of these ideas might be useful to you too.

1. **Breathe.** Breathing is probably the greenest thing any of us can do. Focusing on the breath allows us to take a break from whatever is going on in our heads and just experience the moment. A moment in which we are alive on this beautiful planet that we love. A moment in which we don't need anything. Sitting quietly and meditating for a few minutes each morning is the best way to prepare for what lies ahead. And you know what? Sometimes, breathing leads to crying, and that's okay too. I find that tears can help flush out my worries better than any drug, and crying opens the heart to whatever comes next.

2. **Spend time outside.** Whether I've been sitting at the computer for too long or find myself standing in the middle of the grocery store freaking out about all the plastic I see, stepping outside for a walk always helps. Being in a place with dirt, trees, water, rocks, helps reconnect me to the earth and reminds me why I am doing this work in the first place. But even simply walking around the block and noticing my feet on the ground and the temperature of the air can be enough to clear my head.

3. **Spend time with those I love.** Connecting with other people can help us gain a fresh perspective. Children and animals don't let us take ourselves too seriously. Personally, I can get so involved in the work I'm doing that I neglect my relationships with the living, breathing beings in my life. In working to create a healthier, less toxic world, let's not lose sight of those for whom we do it!

4. **Read inspiring stories.** When I was training for my marathon, I tacked an article about Helen Klein, an 83-year-old ultramarathon runner, to the wall above my desk. Her spirit and dedication motivated me whenever my resolve began to flag. Now, I think of people like Jean Hill, who despite setbacks continued to push her city to ban bottled water. I'm inspired by people who face obstacles and keep going. If they can do it, so can I.

5. **Lose yourself in music or art.** Listening to music I love takes me out of my head and into my body, to a place where my worries and fears can be set aside for a while. As I mentioned in Chapter 3, bands like Radiohead do it for me. And the louder the better. You might prefer something a bit more soothing. Or to engage in singing, dance, or visual arts. It's nice to remind ourselves that humans are capable of creating great beauty in this world, not only destruction.

6. **Write it down.** Sometimes I get overwhelmed simply because my brain is too full of competing ideas. Whether I'm stressing out about the problems in the world or worrying about all the things I have to do, dumping my thoughts onto a piece of paper can help me clear my head and focus on what's most important right now. Then, I can take a break, knowing that the list will still be there when I'm ready to get back to work.

7. **Ask myself what one thing I can do right now.** Getting caught up in how much work there is to do can be paralyzing. From the time I was a teenager, I remember spending entire Saturdays lying in bed thinking about all the things I could do and unable to get up and get started on any of them. Figuring out how to de-plasticize our lives can feel just as paralyzing. That's why it's important to start slow and choose one task at a time. When overwhelming feelings come, it can help to ask ourselves what is the one easy thing we can do right now. And then just do that one thing.

8. **Congratulate myself for my progress.** Instead of obsessing about what I haven't done, it's crucial for me to acknowledge the stage I'm in now. None of us can do more than we can do. In fact, I had to give up thinking I would ever do enough. Seriously. There is always more work to be done, and I don't plan on

ever being finished. But the present moment is all we ever really have. So let's take some time to celebrate how far we've come and who we are in this moment instead of beating ourselves up for not doing more.

9. **Make a game of it.** Seven years ago, the idea of trying to live without plastic was a challenging experiment. It was fun and new and exciting. I made my own rules for the game and invited others to play along if they wanted to. Then, somehow, I lost the fun. Reading article after article about plastic pollution and toxic chemicals left me feeling anxious about the state of our world. I felt compelled to do even more. Writing blog posts, contacting companies, calling my legislators, waging a citizen action campaign, connecting with other organizations and activists, researching issues and alternatives, and still holding down my regular accounting job were taking their toll. I had become as addicted to plastics activism as I had been to plastic itself. Where was the fun? What happened to the game?

One day, in the spring of 2010, I found myself out on Kehoe Beach with artists Richard Lang and Judith Selby, who make art out of found plastic. This particular beach in Northern California is routinely strewn with plastic waste that washes up on the shore with every wave. And we were hunting for treasures, pieces that Richard and Judith could use in their collages and sculptures. We weren't cleaning the beach—there was too much plastic to ever be cleaned up—so much as curating it, looking for interesting objects with which to fill our reusable bags. It was a game. Richard found a plastic baseball bat. I found one half of a plastic Oreo cookie. I can't remember what Judith found, but as I recall, it was just as weirdly cool and disturbing.

Plastic Oreo cookie on Kehoe Beach

And then suddenly, looking at all the plastic embedded in the seaweed, mixed with the sand, tangled in the beach grass, I felt a deep sadness rush over me. It's hopeless to think about cleaning up all the plastic in the ocean. There is no way to reverse the damage we have already done as long as companies continue to produce

> "My strategy to avoid becoming overwhelmed and keep moving forward in my efforts to reduce my plastic is to try just one new change each month. One month I ordered cloth produce bags and stopped using plastic bags. Another month I found plastic-free laundry soap. This pace of change works well for me and keeps me feeling inspired."
> —Tara, Sacramento, California

throw-away plastics and consumers continue to buy them. How can we empty a bathtub with the spigot still running? This problem is just too big!

I froze.

And then just as suddenly, my focus shifted from the plastic stuck in the sand to the incredible blue sky in the distance, and the cliffs behind us, and the sea air on my face, and sadness turned to joy for the beauty of this world we humans both love and destroy. ***Wow,*** I thought, ***just wow.*** Everything looked different. But Judith and Richard were down the beach and apparently hadn't noticed a thing. I got up and kept moving. And I was back in the game.

How do I keep trying when the end result is in doubt? By focusing on the present moment. By realizing that my efforts might not create the result that I think I want but that I have to keep trying because it is who I am. It is the right thing to do. And because living a life respecting the earth and protecting her creatures is something I have to do. Now. No matter what the outcome. Because if we don't try, we are certain to fail.

So every day I re-create the game. I remember the excitement I felt at the beginning of this project when I knew nothing, and every discovery brought me joy. I try things I've never tried before. I challenge myself in ways I never thought possible. I breathe and take walks and connect with the world. And when grief comes, which it sometimes still does, I allow it to wash over me. Because it's real. It's what's here right now. For what more can I ask?

Chapter 9: Durable Goods (When Cheap = Green)

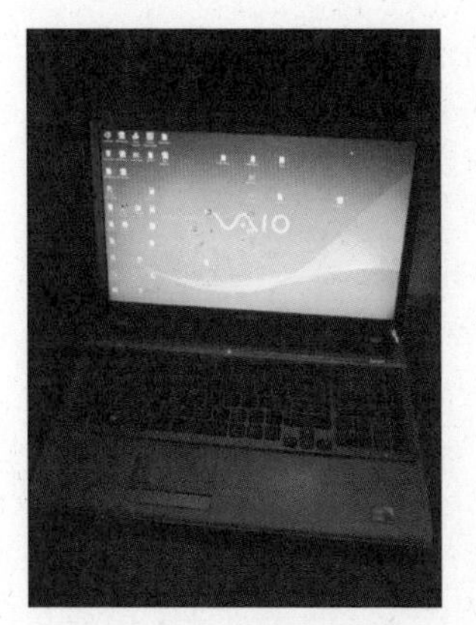

I still use lots of plastic. The computer I'm typing on right now is plastic, as are the polycarbonate eyeglasses on my face and the spandex fibers in some of my clothing. In a few minutes, I'll get up to feed my kitties and clean out their plastic litter boxes. The sky is pretty overcast today, and I'm wondering whether I should take my nylon umbrella with me when I go out tonight.

From the beginning of this project, I sensed that continuing to use the durable plastic products I already owned was more eco-friendly—and wallet-friendly—than chucking it all in favor of plastic-free substitutions. But as I learned more about the toxic chemicals in certain plastics, I decided that some of my plastic possessions had to go: those made from vinyl, for example, or those used for food contact. Still, there's no reason to fill up the landfill with useful plastics that have a lot of life left in them. Here are the steps I've taken to remove the most harmful plastics, continue using those with real benefit, and acquire the few plastic products that truly make my life better without encouraging the production of virgin plastic and without spending a lot of money. Keep in mind that I didn't make all these changes at once and you don't have to either. Set some priorities, be creative, and have some fun.

Get Rid of PVC

As I mentioned in chapter 1, polyvinyl chloride (a.k.a. #3 PVC) is particularly problematic because of additives like phthalates and heavy metals that can not only leach into foods but off-gas into the air and rub off on skin. I decided that my PVC shower curtain had to go, as did vinyl purses, rainwear, mattress covers, and any other soft, flexible plastics with that "new plastic" smell that I suspected to be made of PVC. But go where?

The Center for Health, Environment, & Justice (CHEJ) has examined the problems of PVC disposal and determined that the chemicals in PVC can leach from landfills into groundwater, contribute to dioxin-forming landfill fires, and release toxic emissions in landfill gases. And burning PVC in incinerators is even worse, as it emits toxic dioxins into the air and creates ash that ends up in landfills anyway. CHEJ recommends we collect our PVC products and put them aside in an enclosed cardboard box away from the sun and possible ignition sources. Then, either send them back to the manufacturers, along with a letter urging the companies to switch to safer materials, or cart them off to our local hazardous waste facility.[108] In the beginning of my plastic-free project, I did both of those things: mailing back a #3 squeeze bottle of caramel syrup, a #3 container of sesame seeds, two #3 bottles of hair gel, and even a Multi-Pure water filter unit that I discovered came with PVC tubing (for which I received a full refund after much back and forth correspondence with customer service); and delivering the rest of my PVC possessions to Alameda County's hazardous waste facility for disposal. Sure, the workers there, who were used to collecting old paint, motor oil, pesticides, cleaning chemicals, batteries, and fluorescent tubes, were confused when I handed them my old shower curtain and purses, but what else was I going to do with them?

PVC may be found in a lot of kids' products and school supplies, including toys, teethers, playground balls, lunch boxes, backpacks, plastic-covered 3-ring binders, colored paper clips, raincoats, shiny umbrellas, and polymer modeling clays like Fimo and Sculpey. PVC can be found elsewhere in the home in products like the aforementioned shower curtain, as well as some cling wraps, mini blinds, mattress protectors, plastic table cloths, imitation leather furniture, inflatable furniture and toys, artificial Christmas trees, electronics, wires and cables, carpet undersides, flooring, wall coverings, window frames, doors, and siding. How can you find alternatives?

CHEJ provides two very useful guides to finding PVC-free alternatives. You can download both the ***Back-to-School Guide to PVC-Free School Supplies*** and ***Pass up the Poison Plastic: the PVC-Free Guide for Your Family and Home*** from CHEJ.org.

Phase Out Plastic Foodware

While I was figuring out how to get rid of all the single-use plastics in my life, I continued to use my Tupperware and other plastic food containers and plastic-coated cookware. But slowly I found ways to replace all of it so I wasn't eating from or cooking with plastic.

Set Your Priorities

Here are some questions that can help you decide what your own priorities are and what plastics to focus on first.

1. **What Are the Most Toxic Plastics in My Kitchen?** Those that contain BPA, PVC, or are coated with non-stick surfaces like Teflon are the most likely to leach toxic chemicals. Check The Soft Landing website for guides to PVC-free and BPA-free products. But do keep in mind that even BPA-free plastics may not be entirely safe.

2. **What Dishes Are Used by Children?** Children's bodies are more susceptible to harm from chemicals in plastic. Consider replacing those first.

3. **What Dishes and Utensils Do I Use for Hot Foods?** Chemicals in plastics leach more quickly when heated. Switch to glass bowls and containers for microwaving; non-coated pots and pans and wood or metal utensils for cooking; glass, stainless steel, or cast iron bakeware; and stainless steel, glass, and ceramics for storing hot leftovers.

Mason jars are great to reuse for food storage.

4. **What Dishes Are Used to Hold Greasy/Fatty Food?** Because plastic attracts oily substances, it's best not to store greasy foods in them.

5. **What Dishes Do We Use Most Often?** While it might not be feasible to replace everything at once, think about designating a few nonplastic plates, cups, and containers to use on a regular basis.

Never heat plastic in the oven or microwave, even products that are labeled "microwave safe." A study conducted by the Milwaukee Journal Sentinal found harmful levels of BPA leaching from all of the "microwave safe" containers they tested. BPA expert Dr. Frederick vom Saal of the University of Missouri says, "There is no such thing as safe microwaveable plastic."[109]

Use What You Already Have

If it seems like replacing the plasticware in your kitchen is going to be expensive, remember that the most eco-friendly alternatives are also the least costly. Before buying anything new, shop your own cupboards to discover what plastic-free alternatives you might already own. Ceramic mugs, for example, might substitute for bowls in a pinch. Ceramic bowls and saucers can be used to store leftovers. Glass mason jars can be reused for drinking or for storing almost anything. I'll say more about using glass jars later in this chapter.

Look for Secondhand Kitchenware

My favorite kitchenware shops are Goodwill and the thrift store down the street. While I did buy two brand new stainless steel pots, courtesy of a generous Christmas gift certificate from a friend, I found all the rest of our cookware secondhand. Via an ad

I placed on Freecycle.org, I not only scored some great cookie sheets but also met a new friend.

And even before starting my plastic-free project, I had obtained all of our plates, bowls, glasses, and utensils secondhand. I was just being cheap. I didn't realize at the time I was also being green.

> "I stopped washing any plastic (containers, dishes, cups) in the dishwasher to prevent chemicals from leaching in the heat. Washing them ALL by hand really gave me a feel for how many I was using. The extra work motivated me to find alternatives and transition away from using so much plastic." —Laura, Minneapolis, Minnesota

Replace Plastic Children's Foodware

I grew up in the 1970s, at the dawn of the plastic age. My mom was grateful for the hard plastic plates, bowls, and cups that made her life easier. So do we have to go back to china dishes and take the risk of kids breaking them? Not necessarily. While many parents have told me that their kids do fine with glass or ceramic dishes, others don't want to take the risk. These day, families have options. Here are a few sources:

Baby Bottles and Sippy Cups The Glass Baby Bottle sells baby bottles made from tempered glass or stainless steel, as well as stainless steel sippy cups, some of which have clear, medical grade silicone nipples and sippy spouts instead of plastic.

Baby Dishes and Utensils Life Without Plastic offers children's plates, bowls, cups, and utensils in stainless steel or naturally-laquered wood. And Mighty Nest sells toddler and kid dishes in stainless steel or Duralex tempered glass, which is touted as two and a half times stronger than regular glass.

Stainless Steel Popsicle Molds Two companies, Onyx Containers and Life Without Plastic, offer unique popsicle molds as an alternative to plastic and silicone. While silicone appears to be a safer material than plastic, I would opt for stainless steel over silicone when possible. I'll explain why later in this chapter.

> "Use your teacups as little bowls. Looking for a replacement for our old plastic bowls, I remembered the teacups that came with my everyday dishes that we never, ever use. My 1-year-old is appropriately careful with them, and she loves that her ice cream dish has a handle."
> —Julie Dike, Virginia

Stainless Steel Ice Cube Trays Many moms like to freeze small portions of baby food in ice cube trays. Both Onyx Containers and Life Without Plastic offer stainless steel versions, which work similar to the aluminum ice cube trays of the past but are made from a healthier metal.

Replace Non-Stick Cookware Like Teflon

Stainless Steel is my favorite type of cookware. I've even learned how to fry eggs perfectly in a stainless steel pan. The key is to first heat up the pan, add the butter or oil, and let the egg cook until it develops a brown crust. Trying to turn the egg too early or putting the egg into the pan before the pan is hot are ways to ruin a perfectly good egg. YouTube actually has some great demonstration videos showing how to cook eggs in stainless steel. Another alternative that I learned from my friend Andy is a hybrid fry/poach method. Heat the pan, add the oil, add the egg, and let it start to set up. Then, add water to the bottom of the pan and cover for a minute or so. The water will steam the egg done, so you don't even have to turn it over in the pan.

Note: People with nickel allergies should avoid stainless steel. Other great alternatives for cookware are:

Cast Iron Cast iron can add a little iron to your diet, which is a good thing for most people, and food won't stick if your cast iron is well-maintained and seasoned properly.

Enameled Cast Iron, Like Le Creuset The enamel coating prevents acidic foods like tomato sauce from reacting with the cast iron.

Tempered Glass Glass is the most nonreactive type of cookware, but it can shatter when subjected to extreme differences in temperature or allowed to boil dry.

What About the New Nano-ceramic Coatings? At this point, I think it's best to avoid pans with "nano-ceramic" coatings, as we just don't know enough about the potential impacts of nanotechnology at this time. Even the EPA has stated that nanotechnology is a relatively new science, and as a result, the health implications associated with engineered nanomaterials have not been determined. Much research is still needed.[110]

Whichever Type of Cookware you choose, when buying new (as opposed to secondhand), look for plastic-free options made without plastic handles or other trimming. There are all-metal or metal/glass options without any plastic at all. Why buy new plastic if we don't have to?

"We use regular ceramic dishes for kids. We bought a stack of salad plates at Goodwill, so we weren't out much money if they got broken. Four years later we still have most of them. They were light enough that my 1 and 5 year olds were able to bring their own plates in from the table. Kids seem to naturally be more careful when carrying the heavy weight of a glass or ceramic dish than they are carrying light-weight plastic." —Melody, Santa Cruz, California

Switch to Plastic-Free Dishes and Food Storage Containers

We store foods in the cupboard, refrigerator, and freezer in several different ways, depending on how quickly the food will be consumed.

Cloth Bags and Wraps Abeego makes reusable cloth wraps infused with beeswax and tree resin as an alternative to plastic cling wrap. Many kinds of produce do well in cloth bags or bowls covered with cloth towels or wraps. See the list in chapter 6 for more information about how to store produce without plastic.

Cereal Bowl with Saucer On Top I learned to store food in a bowl with a saucer on top from Michael's mom, who has never liked plastic wrap. This method is most appropriate for refrigerating leftovers that will be eaten quickly because the saucer is not an airtight lid. The nice thing about this solution is that the bowls and saucers are stackable.

"I use a tiny stainless demitasse spoon to feed my baby since all other baby spoons are coated in plastic. And I don't feed her from a plastic bowl but just from a coffee mug." —Olivia Mather, Long Beach, California

Repurposed Glass Food Jars Glass jars are too valuable for the recycle bin. Our cupboards, refrigerator, and freezer are full of all different sizes of glass jars that we use to store cut avocados, limes, and lemons; homemade pesto and hummus; nuts, trail mix, rice, and flour (nuts and whole grains stay fresher longer in the refrigerator); beans; pasta; bulk oils and nut butters; dried fruit; grated cheese; and whatever else will fit in a jar. Freezing in glass jars is fine as long as you follow a few simple rules: Leave an inch or two of space at the top of the jar, as the food will expand when frozen. And do not subject the jar to extreme temperature differences, going from freezer to microwave, for example, or running a frozen jar under hot water.

"I freeze my homemade tomato sauces in mason jars with the tops still loose and then close them once the sauce has frozen." —Annie Urban, www.phdinparenting.com, Ottawa, Ontario, Canada

Weck Canning Jars If you're concerned about the BPA in metal jar lids, consider investing in some Weck canning jars, which come with glass lids and natural rubber gaskets.

Anchor Glass Refrigerator Containers Anchor glass makes several sizes of plastic-free, stackable refrigerator containers with flat glass lids that sit on top. The containers are not airtight, so they are best for foods that will be eaten fairly quickly. The pint-sized containers are the perfect size for freezing one day's serving of our homemade cat food, which will be consumed within a week or two. But they are not appropriate for long-term food storage.

Airtight Stainless Steel and Glass Containers Life Without Plastic offers a variety of plastic-free airtight containers in both glass and stainless steel. Both containers have metal lids and employ a silicone gasket to create the seal. These kinds of containers are perfect for long-term freezer storage of just about any food, including garden produce, baked goods, and even meat.

Freeze Foods without Ziploc Bags If you've followed the steps in Chapter 6 for storing fruits and vegetables without plastic and find that they still don't last long enough, you might be like me. I do eat kale. I just don't eat it fast enough to keep the last few leaves from turning yellow in the refrigerator. I thought about freezing my greens, but how could I freeze individual portions without Ziploc bags? Wouldn't it all get hard and stick together?

Actually, freezing greens and other veggies without plastic is quite easy. To freeze them raw, just wash, spin dry (see Page 283 for my plastic-free salad spinning discovery), and store in an airtight stainless steel or glass container in the freezer. As long as you've removed most of the water, you can easily break off pieces when you're ready to eat them. Or, if you'd rather store cooked greens, you can steam them, drain off any excess liquid, and freeze in individual-sized clumps on a flat tray or plate. Once the portions are hard, toss them all into your airtight container. They won't stick together. I love to add a handful of frozen veggies to my daily morning smoothie. No need for ice!

A freezer full of glass and stainless steel.

Learn to Preserve Foods without Freezing Some people avoid the whole question of what kinds of containers to use in the freezer by learning to can produce in glass jars or even dry fruits and vegetables using a dehydrator or a low temperature oven. Visit the website of the National Center for Home Food Preservation to get started.

Don't Forget Pets Michael brought me the strangest story a while back. His workmate's cat had developed chin acne, and her vet told her to stop feeding him in plastic bowls. Even I thought that sounded strange, so I Googled "cat acne plastic," and sure enough, the Internet was full of anecdotes about cats and dogs developing acne from eating out of plastic bowls. Why would this happen? One theory suggests that, just like plastic cutting boards, plastic bowls are more difficult to clean and thus allow acne-causing bacteria to breed.[111] It could also be that some cats and dogs are allergic to plastic. Whatever the reason, it's best to feed our pets out of glass or metal, just like we feed ourselves.

De-Plasticize Your Appliances and Other Kitchenware

Once we get the plastic out of our cookware and dishes, it's time to look at other plastics in contact with our food.

Food Processors and Blenders The pitcher in many food processors and blenders is made of Lexan, a brand name of polycarbonate plastic, which contains BPA. Newer BPA-free plastic blender pitchers may not be any safer and may pose some of the same health risks as those with BPA. If you use a blender, consider switching to one with a glass pitcher, or try a stainless steel immersion (stick) blender that doesn't allow plastic to come into contact with your food at all. I love my simple Waring Pro blender with its glass pitcher, but if you need a super high speed model, consider searching eBay or thrift stores for a vintage Vitamix. The original Vitamix blenders came with a stainless steel container instead of the plastic ones offered today, and many of them still work great. Jay and Chantal from Life Without Plastic swear by theirs. Unfortunately, all consumer-level food processor pitchers are made from plastic. (There are commercial food processors with metal bowls.) Avoid processing hot foods in them, and don't subject the pitcher to the high heat of the dishwasher. I still use my food processor, which I received as a gift many years ago, to make pesto and hummus and to shred cheese. Personally, I'm willing to risk the small amount of BPA because the food isn't hot and doesn't stay in the food processor for very long. You may feel differently and want to avoid even that small amount of BPA.

Coffee Makers It doesn't make sense to pour hot coffee into a plastic cone. Fortunately, there are plastic-free options. Melitta offers a porcelain pour through coffee cone, and Chemex offers a glass option. Add a reusable cotton or hemp coffee filter to avoid paper waste. For those who prefer the flavor of coffee made with a French press, Frieling makes a double-walled insulated all stainless steel model. As for me, ever since I discovered the cold brewed coffee fad, I haven't needed any coffee maker. I combine coffee and water in a mason jar, let it sit on the counter until the next morning, and then strain it through a sieve. The taste of cold brew is mellower than hot coffee and super easy for a lazy person like me.

But speaking of fads, there's another recent fad that makes my head hurt just to think about it: the pods. The name sounds like something from a horror movie, and considering their environmental impact, it should. Coffee pods like Keurig's K-cups and Nespresso capsules, whether recyclable or not, are single-serving, single-use disposable products that take a tremendous amount of energy and materials to produce and ship. If you already have a single-serve coffee brewing machine, consider investing in a refillable stainless steel coffee capsule, and fill it with your own coffee. MyCoffeeStar capsules fit Nespresso machines and the Ekobrew Elite capsule fits most Keurig models.

Choose a Wooden Cutting Board While many people assume plastic cutting boards are more hygienic than wooden ones, the opposite is actually true. In a UC Davis study comparing scratched up wooden and plastic cutting boards, the wooden boards were found to be naturally antimicrobial, while plastic cutting boards allowed more bacteria to thrive, even after being washed.[112] In an effort to protect plastic cutting boards from microbes, manufacturers nowadays treat plastic cutting boards with antibacterial and fungicidal chemicals. I'd rather skip the chemicals and use the board that is naturally resistant.

Searching for a Salad Spinner One afternoon, sick of my plastic salad spinner that was hard to clean and took up way too much valuable counter space, I went on a Google hunt to find a plastic-free salad spinner. There had to be one, but even those ad-

A tale of two plastic-free salad spinners

vertised as stainless steel turned out to be plastic on the inside. Somehow, I found myself on Youtube watching an old Julia Child episode from the 1960s: "How to Make Salade Nicoise." My jaw dropped around minute 10:55 when Julia pulls out her French salad spinner: a wire mesh basket that spins like a top in the sink! So of course, I went hunting on eBay and Etsy to find a vintage salad spinner. I never found the spinning top one, but I did find one that you whip around over your head like a windmill . . . outside the back door or in the shower if you don't want to splatter your kitchen. I bragged about my find on Facebook, but several friends took the wind out of my sails, asking why I didn't just use a tea towel or old pillow case. Huh, really? So I did an experiment: wire basket vs. cloth produce bag (I figured the bag would be better than a pillow case or towel because it closes shut), and you know what? They were right! The cloth not only wicks the water away from the salad greens, it also absorbs any water that is left. And whipping that bag around is good exercise for you and a good laugh for your friends. Everybody wins.

The East Bay Depot for Creative Reuse collects unwanted plastic containers for teachers and artists to use.

What Do We Do with Old Plastic Kitchenware?

In our home, we still use many of our old plastic containers and bottles, just not for food. In addition to storing our No 'Poo hair care mixtures in plastic sports bottles in the shower, we use an old Tupperware container to scoop cat litter out of the bag and another one to hold the various lids for our growing glass jar collection.

Throughout our house, plastic containers hold cleaning supplies, rags, first aid supplies, hair accessories, buttons, sewing notions, colored pencils, paper clips, nails and screws, and all manner of small objects. I asked my blog readers what they do with old plastic foodware to avoid sending it to the landfill and got a lot of creative suggestions:

- Organize toys, puzzle and game pieces, dice, crayons, rubber stamps, cookie cutters and other kitchen gadgets, coins, ribbons and bows, cosmetics, jewelry, tools, fishing tackle, or used batteries waiting to be recycled in plastic cups, bowls, and containers.
- Use plastic bowls, containers, cups, and utensils as shovels and buckets for scooping and digging in the sandbox and as molds for making sand castles. (A sandbox is a great place for old plastic, but please avoid bringing plastic to the beach since it can so easily be washed into the ocean.)
- Punch drainage holes in the bottoms of plastic bowls and containers and use them to plant seedlings. Place plastic lids under plant pots to catch water.
- Mix paints in plastic cups, bowls, or containers and use plastic lids as reusable palettes.
- Keep homemade playdough in airtight plastic containers.
- Turn plastic foodware into musical toys. Fill lidded containers with dry beans to use as shakers. Transform larger containers into drums.
- Make a small slit in a plastic container to use as a piggy bank.
- Fill containers with water and freeze to use as ice packs in insulated chests or to take up unused space in the freezer for greater energy efficiency.

Those of us who have amassed huge collections of plasticware may run out of personal uses for it all. Consider donating items in good condition to classroom art teachers, play groups, day care centers, animal shelters, garden centers, farm stands, or co-ops for reuse. Offer plastic items and old cookware on Freecycle.org or give to local thrift shops. Some of my blog readers have wondered about the ethics of passing off our unwanted plastic to people who might use it for food contact. My philosophy is that giving it to someone who would otherwise have purchased brand-new plasticware reduces the demand for new virgin plastic to be produced. And that's a good thing for all of us.

Plastic that is too worn out to be reused should be recycled if possible. Check and see if there is a resin code number on the bottom and if your local recycling center will accept it. Scrap metal recyclers will sometimes accept old cookware, but check first to find out if they will recycle pots and pans with Teflon coating. Use the Earth911 website to locate and contact recyclers in your area.

What about Silicone?

I've wondered about silicone as an alternative to plastic for a long time. Some websites claim that it's a natural substance and 100 percent safe. Others worry about leaching from silicone products. To get to the bottom of the silicone mystery, I consulted Joseph Krumpfer, a polymer chemist studying silicones at the University of Massachusetts. He helped me get a better understanding of this material.

First of all, silicone is no more "natural" than fossil-based plastic. It is a man-made polymer, but instead of a carbon backbone like plastic, it has a backbone of silicon and oxygen. (Note that I'm using two different words here: silicone is the polymer and silicon, spelled without the "e" on the end, is an ingredient in silicone.) Silicon is an element found in silica, i.e., sand, one of the most common materials on earth. However, to make silicone, silicon is extracted from silica (it rarely exists by itself in nature) and passed through hydrocarbons to create a new polymer with an inorganic silicon-oxygen backbone and carbon-based side groups. What that means is that while the silicon might come from a relatively benign and plentiful resource like sand, the hydrocarbons in silicone come from fossil sources like petroleum or natural gas. So silicone is kind of a hybrid material.

Is silicone safer than plastic? At this point, most experts believe it is. Silicone has a much higher heat stability than carbon-based plastics and is less likely to leach toxic chemicals. But the truth is that not as much research has been done on silicones as on plastics. According to *Scientific American*, the FDA determined back in 1979 that silicon-dioxide, the raw material for silicone, was safe for food-grade uses, but since the first silicone cookware began to show up on store shelves a decade later, no follow-up studies have been done to determine whether chemicals can leach out of silicone cookware and contaminate food.[113] While I believe that clear, "medical-grade" silicone is most likely the safest material for baby bottle nipples and sippy spouts, and small silicone rings and gaskets are appropriate for sealing glass and metal containers, I wonder about all the hot pink and ultramarine toys and kitchenware made from "food-grade" silicone. What additives are in the polymer? What ill effects might we discover down the road? Personally, I've decided not to use silicone cookware until more research is done proving that it is safe. After all, we thought that Teflon was safe for decades.

Whether or not silicone is safe for food contact, there are other environmental impacts to consider. In addition to the fact that some of its components come from fossil sources, silicone may not be any more biodegradable than plastic. What happens to silicone consumer products that get loose in the environment? How will they affect the ocean and marine life? My feeling is that we should exercise the Precautionary Principle and limit our use of this complex material until all the facts are known.

Using Plastic Responsibly

Life is full of unexpected mishaps and happy surprises. How do you refrain from buying new plastic when your computer monitor dies? Your rice cooker stops working? Your umbrella self-destructs in the middle of a thunderstorm? How do you cope when you suddenly find yourself the proud parent of two adorable kittens and realize you'll need a few plastic items you hadn't considered before, like cat carrier boxes and litter pans? How can you enjoy the benefits of modern society and its amazing technological innovations without succumbing to consumerism and the addiction to buying new stuff? I've had a blast figuring out solutions to these questions, and I've come up with a few basic strategies to avoid buying new plastic while still obtaining the benefits of the plastic products that truly make my life happier.

Fixing Things

Learning to fix things when they break has been both frustrating and rewarding. Frustrating, because companies nowadays produce most products to be replaced rather than repaired when they inevitably fall apart. Rewarding, because repairing something myself gives me a sense of freedom and self-sufficiency akin to having a super power. When one of the metal ribs on my umbrella snapped, I fixed it with a paper clip and a pair of pliers. Instead of tossing a cracked plastic laundry basket, I followed the lead of blog reader Ellen Simpson (who fixed her laundry basket with a piece of broken umbrella) and reinforced the rim with a wire coat hanger. I've enlisted the help of experts like my dad—a retired electronics technician—and various handy people on Craigslist to tackle a burned-out hair dryer, dead rice cooker, and digital scale. And I even took apart and climbed inside my smelly washing machine in an effort at gross defunkification.

I report all my repair successes on my blog as a way to add to the growing body of do-it-yourself knowledge on the Internet. Sites like ifixit.com, fixya.com, Instruc tables.com, eHow.com, HowToDoThings.com, applianceaid.com, appliantology.org, and fixitnow.com are invaluable resources. Youtube is full of great "How To" videos. And Craigslist.org is a fantastic tool for finding local people with the hands-on experience to help you out.

Upgrading software and/or components is a way to keep computers working longer instead of buying new ones. At my office, we purchased a new computer program, only to discover that our old computers would not run it because of their ancient operating systems. Being the office computer geek, I saved the company lots of money (and saved the environment from all the toxic chemicals in the old machines) by learning how to reformat the hard drives, install a new operating system, and reinstall all the old programs. The computers worked faster and better after their upgrades, and I went home feeling like a hero. I've also learned to install memory cards and replace hard drives and power supplies, all of which consume fewer resources than a whole new computer.

But what about things that's can't be repaired or upgraded? Sadly, many companies follow a policy called ***planned obsolescence***, or as environmental author Annie Leonard calls it, "design for the dump," deliberately designing products with a limited useful life so that they will become nonfunctional after a certain period of time. It's a way to force customers to buy new products on a regular basis instead of repairing or upgrading what they already have. I learned all about planned obsolescence one morning a few years ago when I turned on my Hewlett Packard computer as usual, pressed the button on my monitor, and nothing happened. Actually, the computer came on—I could hear its happy hard drive spinning up—but the monitor did not. I made sure all the cables were connected, plugged it into different outlets, and stared at a black screen. Not even a blink. I called HP and was told that the monitor, which was only one and a half years old, was out of warranty, but they'd be happy to sell me a new one.

No way was I going to replace such a new piece of equipment! So I got out my handy Yellow Pages (since I couldn't use my computer) and found a local computer repair guy named Leon who said he'd be happy to take a look at the monitor. He took it apart, tested the circuits, and determined that we needed to replace one small board inside the machine. But when he called HP to order the part, he was told that the company does not sell replacement boards for these monitors and that our only recourse was to send the monitor back to HP for recycling and buy a new one. What's more, the board was a proprietary part that could not be obtained through any other source, so there was no way for Leon to fix it, even though he knew how.

Designing gadgets that can't be serviced, with proprietary parts that can't be replaced, is one way companies keep us on the consumer treadmill. Another way is through developing new technologies while still manufacturing and selling old ones, planning for the old gadgets to become obsolete as soon as the new ones come out. And restyling the external appearance of gadgets is a way to increase sales by appealing to consumers' desire to be seen as hip and cutting edge. For example, ten months after releasing its iPhone 4, Apple came out with a white version of the exact same phone and created such a mystique around the product that even many customers who already owned the original black version purchased a white one simply for its looks. What's more, Apple knew at the time it released the white iPhone that it would be obsolete in six months when the next version came out.[114]

We don't have to be sucked in by marketing hype. And we can refuse to allow our things to become obsolete by daring to open the case and tinker with what's inside. We can own the things we buy, instead of allowing our things, and the manufactures of those things, to own us. Check out IFixIt.com's Self-Repair Manifesto, and insist that companies recognize our right to repair our own things and provide us with the tools, parts, and instructions we need to do it.

Buying Secondhand Plastic

So, what did I do when I couldn't repair my HP monitor? Cave in and buy a new one from HP? No way! I checked Craigslist.org and found a secondhand Dell monitor for less money. The used monitor has a tiny scratch in the upper left hand corner, which I never notice when I'm working. I was happy to give this imperfect but perfectly functional gadget—which to date has outlived its predecessor by several years and is still going strong—a new home.

Even before I stopped buying new plastic, I was an avid thrift shopper and eBayer. I wasn't necessarily being green, just trying to save money. But the truth is that some secondhand things are actually higher quality than the cheap plastic crap churned out today. A vintage KitchenAid mixer, for example, can last for generations and is built with parts that can be repaired. The wedding outfit I purchased from an eBay seller—a vintage green and gold I. Magnin suit from the 1960s with matching hat—was stunning and original and cost much less than a new dress with less character.

Since going plastic-free, we've continued to acquire the things we need secondhand. We got our cat litter boxes from kind, newly cat-free people via Freecycle and found our cat carrier boxes at a couple of secondhand stores. We had heard that cats might reject litter boxes used previously by other kitties, but Soots and Arya had no trouble using well-scrubbed hand-me-downs. When I decided to give crock pot cooking a try, I found a perfectly good one for $10 on Craigslist and didn't have to worry about whether I would use the machine enough to make up for all the resources used to manufacture it. I didn't buy it new.

In fact, I may have saved it from the dump. I do most of my clothes shopping at consignment shops and thrift stores in my neighborhood and even found the perfect birthday present for a coworker secondhand. Shhh . . . she doesn't know. My biggest score is the laptop computer I used to write this book. When Michael's ancient desktop computer finally bit the dust, I gave him my HP and went in search of a used laptop that could do double duty as my main computer as well as travelling companion for presentations and conferences. It took a while to find, but unwilling to purchase a brand new machine, I waited until I came across the perfect laptop at a secondhand electronics shop in Berkeley. Buying secondhand keeps other people's unwanted items out of the landfill, reduces demand for new plastic products, and saves a ton of money.

> "I go shopping at thrift stores and yard sales if I need anything 'new.' Not only do I get the clothes and household goods I need without any plastic packaging, but I end up buying better quality fabric and furniture to boot!" —Chris T, www.simplesavvy.wordpress.com, Belmont, Massachusetts

Sources of secondhand items:

- Yard sales and estate sales
- Community or neighborhood swap meets
- Thrift stores, vintage shops, and consignment shops
- Antique shops and secondhand furniture stores
- Freecycle.org. Sign up to give away or receive other people's unwanted items for free within your local community.
- Buy Nothing Project. Join a hyper-local Facebook group to give away or receive gifts from your nearest neighbors. No group yet in your area? Create one!
- Craigslist.org. Find local people selling what you need or list your used items for sale.
- Yerdle. Download this iPhone app to purchase secondhand goods with Yerdle credits and to earn Yerdle credits by posting your own used stuff for sale.
- The Reuse Alliance promotes reuse over recycling and provides a helpful list of local exchanges for scrap and salvaged materials to reuse or repurpose. Check their membership list for resources near you.
- eBay.com. Bid for items from individual sellers or auction off your own treasures.
- Listia.com. Bid for items or auction your own using credits instead of money.
- Swap.com (formerly Swaptree.com). Buy, sell, or swap pre-owned books, music, movies, games, clothing, toys, baby and child gear.

- SwapMamas.com. Swap clothing, toys, baby and child gear, household items, coupons, and more with local or long distance swappers. The giver always pays for shipping.
- Sister sites PaperBackSwap.com, SwapaCD.com, and SwapaDVD.com. Swap books, music, and movies.
- GearTrade.com. Buy and sell used outdoor gear.

Note: When using any online resources to sell or swap items through the mail, make sure and ask the seller or giver to send items without plastic packaging.

Buy Refurbished

Another way to buy secondhand items is to opt for refurbished or remanufactured versions of gadgets instead of buying brand new. When I switched mobile phone companies a while back, I had to buy a new phone, since the technology in my old phone was not compatible with that of the new service. Frustrated, I ranted a bit to the customer service rep, letting her know that I didn't feel good about replacing my mobile phone when it still worked just fine and that even though I wouldn't have to pay extra for the new phone, it was still a big waste of planetary resources. She mentioned that the company had a few reconditioned phones available, which had been used and traded in by other customers, and she could offer me one of those instead of a new one. I felt better about that choice. A reconditioned phone will probably not be the latest model, but the one I chose worked perfectly well for my purposes and was actually more advanced than the phone I already had. It never hurts to ask. Here are a few other sources of refurbished/reconditioned electronics:

Local Used-Electronics Stores Are there used-electronics stores in your neighborhood that take back and recondition pre-owned electronics? Do they offer a warranty on the products they sell? If so, consider patronizing them first. My laptop came from the local CeX store in Berkeley. Founded in London in 1992, CeX now has stores in many countries around the world, buying and selling secondhand electronics from the local community. By buying locally, I avoided transportation fuel and packaging waste and supported the local economy.

Manufacturers' Websites Many electronics manufacturers sell refurbished and reconditioned equipment right through their own websites. Check and make sure the product you're interested in comes with a warranty.

Online Retail Outlets Outlets like Amazon.com, Buy.com, Overstock.com, and many other online electronics outlets offer refurbished gadgets. Make sure you read all the fine print before purchasing.

Green Citizen To help with electronic waste, Green Citizen refurbishes and sells pre-owned electronic equipment and parts via its eBay and Amazon stores, and responsibly recycles anything that can't be repaired.

Make sure to ask for no plastic packaging if you order online.

Borrow/Share/Rent

Out here in Oakland, California, the grass really only grows tall in the spring. As renters, Michael and I had always relied on our landlord to send someone once a year to mow our overgrown front yard, but in the spring of 2009, I got tired of waiting and decided to do it myself. The thing was, I didn't own a mower. So I started checking Craigslist and all the usual sources for used hand-reel lawn mowers (the simple motor-less kind) until Michael stopped me and asked, "Beth, how often do we ever have to mow? You're going to buy a mower for the one time a year when we actually have rain and the plants grow by themselves? Where will we store it?" The man had a point. I started to think about all the garages in the neighborhood storing all the mowers and other tools that only get used a few times a year. It didn't make sense. We could save so much money and so many resources by sharing tools, appliances, sporting goods, and other things that we use infrequently.

That spring, I discovered the Temescal Tool Lending Library, part of the Oakland Public Library system, which loaned me a hand mower for free! I chuckled to myself as I wheeled it down the sidewalk back to my house, imagining people wondering why I was mowing the cement. I didn't care. It was a beautiful day, and I was getting some exercise. Two years later, I made use of the Tool Library again while installing my little veggie garden in the same front yard. I needed a wheelbarrow for moving the dirt that had been delivered and dumped onto the sidewalk. Thinking about where I could find a secondhand wheelbarrow, I realized that I was once again looking at the problem all wrong. I didn't need a wheelbarrow—I needed to move the dirt. Borrowing allowed me to accomplish the task without being stuck with a tool that I'd likely never use again.

So far, only a few cities have official tool lending libraries, but a sharing economy is sprouting up all over the place. Neighbors get together to pool their resources informally, swap clothing and toys, share tools and appliances, and get to know each other in the process. Renting is another way to access the functionality of an item without having to buy it and store it. It turns out, you can rent almost anything these days! Here is a list of helpful online resources for borrowing, sharing, and renting:

Borrow Tools from a Local Tool Lending Library. LocalTools.org maintains a list of existing tool lending libraries throughout North America and has great resources for starting a tool lending library in your community.

Borrow or Rent from Your Neighbors. Sign up with online community marketplaces that facilitate person-to-person rental and sharing. Rent and/or share with neighbors and friends through sites like Streetbank or NeighborGoods.

Rent from Individuals or Businesses. Check out websites that help you find anything you want to rent, whether it's available through rental businesses or individuals. Zilok is one example.

Rent Outdoor Sports Gear, including tents, backpacks, sleeping bags, and other items that are usually made out of synthetic materials. Renting is a good option if you don't camp or hike very often. Check out the rental offerings through REI. LowerGear and Outdoors Geek rent gear and also sell secondhand items.

Rent Formal Wear. What if you had access to a dream wardrobe from major designers for black tie galas, weddings, dates, and holidays at 90 percent off retail prices? You do. Rent the Runway lets you rent designer dresses and accessories for formal occasions through the mail.

Borrow, Rent, or Stream Movies Instead of Buying Dvds. These days, I think long and hard before adding to my DVD collection. CDs and DVDs are made from polycarbonate plastic. The only movies I buy are independent educational films that are not available to rent. Otherwise, I rent or stream through Netflix, Google, Hulu, or other online movie sites or borrow from the local library.

Join a Bike Sharing Program. In cities around the world, bike sharing is becoming more and more popular as a quick way to get from one place to another within a small area. Click on the map at The Bike-Sharing Blog to find out about current and upcoming bike-sharing programs.

Join a Car Sharing Program. Give up the hassle of car payments, insurance, registration, smog checks, parking fees, and all the other headaches that go along with car ownership. Some people use their car share membership as their second car. Others, like Michael and me, use car sharing as our only car and opt for public transit and bike riding as our primary means of transportation. We belong to Zip Car. Check out CarSharing.net to find a car share organization or company near you and to read more about car sharing.

Learn More from Shareable.net. Shareable is a nonprofit online magazine reporting on people and projects creating a sharing economy. Check out Shareable's "How to" page for a complete list of How To posts on how to share, rent, swap, or barter just about anything.

Note: As with buying secondhand goods, when using any online resources to share or rent items through the mail, be sure to request the sender use no plastic packaging.

Buy Products Made from Recycled Materials

Recycled plastic is not my first choice for most products. Before going that route, I'd rather refurbish what I already have or look for secondhand options. Repair and reuse have less environmental impact than melting plastic down into pellets to make into secondary products. But sometimes those choices are not available and plastic really is the best material for the job. Our Urban Compost Tumbler, for example, was made from 100 percent food-grade recycled plastic.

When choosing products made from recycled content, it's important to note whether the recycled material used is "preconsumer" or "postconsumer" waste. Pre-consumer content is manufacturing scraps that are recovered and reused in the factory. This kind of waste has never actually had a first life as a consumer product. It makes economic and environmental sense for companies to use this material rather than throwing it away, but to me, this process is not true recycling. Postconsumer material has already had a first life as something else before it was made into your jacket or umbrella or carpet. It could have started out as a plastic bag or plastic bottle. Choose recycled products made with the highest amount of *post*consumer content possible.

Here are a few interesting recycled plastic products I've discovered:

Vacuum Cleaners Electrolux is designing some of its newer vacuum cleaners with 55 percent postconsumer plastic (and higher percentages in Europe.) Type "recycled" into the Electrolux website's search bar to find models that contain recycled content.

Umbrellas The Shedrain Ecoverse umbrella has a canopy made of 100 percent recycled PET soda bottles and either a bamboo or recycled rubber handle. Monsoon Vermont sells umbrellas made from upcycled plastic packaging.

Rainwear Patagonia incorporates recycled material into many of its items and takes back all of its products to be recycled. Type "recycled" into Patagonia's search bar to find products made with recycled content. Kamik makes synthetic rubber rain boots that can be returned to the company to be recycled into black rain boots. (Choose black boots for the highest percentage of recycled content.)

Camping/Outdoor Equipment These days, many outdoor supply companies offer camping gear like sleeping bags and pads, backpacks, and tents made with recycled plastic or recycled polyester fabric. Check out Patagonia, The North Face, Marmot, Big Agnes, REI, and others. Offerings change from season to season, so type "recycled" into the company's search box to find out what gear contains the most recycled materials.

Technical Athletic Wear Atayne manufactures technical athletic shirts and jerseys out of 100 percent recycled plastic bottles and polyester and takes back any brand of fully worn polyester athletic shirt to be recycled.

Shoes Several companies are adding recycled and bio-based content to their shoes to make them a little more eco-friendly. Check out Brooks's "Green Silence" shoes. Kigo makes minimalist shoes from recycled content and takes back its shoes for recycling. In fact, the Kigo Leon model is my new favorite walking shoe. As with all clothing, styles and models of shoes change each season. When shopping for new shoes, contact companies to find out which ones are made with the highest percentage of postconsumer content.

Problems with Polar Fleece I mentioned in chapter 4 that a lot of water bottles are recycled into polar fleece. Fleece is lightweight and comfy and might seem to be a reasonable second life for a water bottle, except that a recent study has shown a downside to this material. Every time a garment made from polyester or other synthetic fabric is washed, it sheds tiny plastic fibers, and fleece fabrics shed the most. This synthetic lint is turning up in the ocean, where the micro-plastic particles act like a sponge, absorbing and concentrating fat-soluble pollutants.[115] Perhaps turning plastic into fuzzy, linty fabrics is not the best idea after all.

MAKING IT YOURSELF

What do you do when your iPod cover cracks and you need a new one and all you can find to buy are covers made out of plastic or handmade covers from Etsy that just don't fit your lifestyle and the way you want to use your device? This might seem like a silly concern when the mp3 player itself is made from plastic, but my plastic-free life is not about deprivation but finding creative ways to enjoy the possessions I already have without buying *new* plastic. So one night, when I was supposed to be working on a million other things, I entertained myself by devising my perfect knitted iPod cover, complete with holes for the power cord and headphone jack and openings for viewing the screen and accessing the click wheel. I felt pretty tickled with myself, and even more so a year later when the foam pads on a couple pairs of headphones wore out and I taught myself to crochet in order to make some plastic-free replacements out of wool. I feel a real connection to and appreciation for items I've spent hours making myself. I also feel much more motivated to take care of them properly. Whenever I need something new, I check Google first to see if someone has come up with a handmade version or instructions for doing it myself. I'm not always willing to take the time, but when I do, it's always worth it.

Handmade headphone ear pads

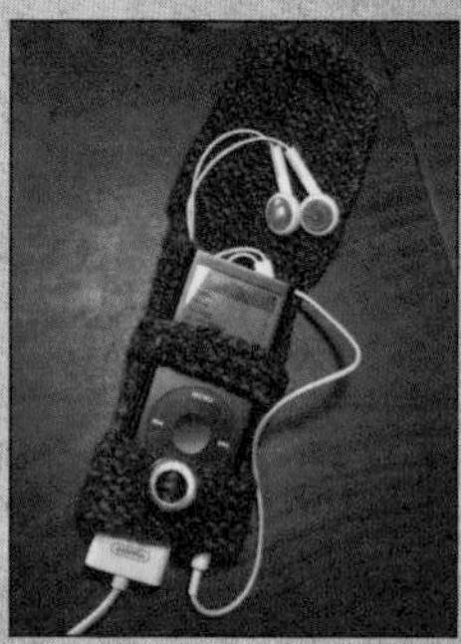

Handmade iPod cover

Buying New Plastic-Free & Less-Plastic Products

Plastic-free products made from natural materials may be more expensive than their cheap plastic counterparts, which is one reason to opt for secondhand items when possible. When buying new clothes, toys, and items for the home, choose quality over quantity. A few beautiful, well-made possessions can last a lifetime or more and may have more meaning than inexpensive junk that breaks readily and is meant to be tossed within a season. Here are some sources for healthier, natural products.

Clothing

I've spent hours scouring clothing racks in retail stores looking for natural fibers, only to be met by polyester, acrylic, lycra, spandex, nylon, and other synthetic materials, i.e., plastics. Even conventional cotton is problematic because of all the pesticides and chemical fertilizers used to grow it. But there are quite a few online sellers of clothing made from natural fabrics like organic cotton, ethically raised wool, hemp, linen, or other plant-based materials. Here are just a few:

- **Stewart + Brown** designs gorgeous women's clothing made from organic cotton, hand-combed Mongolian cashmere, as well as green fibers like Tencel gauze, hemp silk charmeuse, hemp jersey, and linen, which are cultivated without the use of herbicides, pesticides, or defoliants and are sustainably harvested. S+B clothing is expensive, but if you wait until the end of the season, they usually have some great sales. My absolute favorite top is an S+B classic tee that is more flattering than any other piece of clothing I own.
- **Indigenous Designs** produces fair-trade, organic handmade clothing from organic cotton, as well as ethical alpaca, silk, and merino wool.
- **Aventura Clothing** sells lots of organic cotton casual clothing, including my favorite hoodies of all time. What can I say? I like to dress like a five-year-old sometimes. Back in the 1980s, one of my college professors told me I needed to work on my image. I guess I didn't listen.

- **Hempest** offers an assortment of clothing and accessories for men and women made from hemp and organic cotton. Read descriptions to find clothing without synthetic fibers.
- **Faeries Dance** sells a wide array of women's clothing and accessories, as well as some clothing for men and children, in many different eco-friendly fabrics. I like that the site allows you to shop by fabric and lets you know where the clothing is manufactured.
- **Rawganique** offers sweatshop-free, chemical-free, cruelty-free hemp products that are made in the United States, Canada, Europe, and Thailand. Many items are free of synthetic fibers, but you should read the descriptions to make sure. I love that they have organized their bras by how much and what types of synthetic materials are included. It makes shopping for the least plastic products very easy.
- **BeGood Clothing** advertises itself as "Organic clothing at wildly affordable prices." This San Francisco-based company offers classic designs that won't go out of style next year.
- **PACT** underwear is made from 95 percent organic cotton with low impact inks and dyes and is packaged in compostable bags. Unfortunately, the remaining 5 percent is synthetic elastane, which gives the underwear its stretch. While a completely plastic-free pair of underwear would be ideal, I appreciate that a portion of the proceeds from several of PACT's collections are donated to environmental causes.
- **Cottonique** sells 100 percent organic, unbleached, undyed, latex-free, plastic-free cotton socks and underwear for men, women, and children. The clothing is intended for people with chemical sensitivities, so the styles are very plain and simple. The company even offers bras with tie strings intead of elastic.
- **Wild Dill** sells cute organic clothing for babies and small children.
- **Feelgoodz** makes super comfortable flip flops from natural rubber rather than plastic.
- **prAna** As of this writing, the company offers mostly clothing with some synthetic fibers, although recently the company has added more natural fab-

rics. In fact, my favorite running pants are pRana's Cecilia knicker, made from 53 percent Hemp, 42 percent Organic Cotton, and only 5 percent Spandex. But the main reason that prAna is listed here is that, recently, the company made a shift to reducing the plastic bags that their clothing had previously been packaged in. It's a sad truth that most clothing is wrapped in plastic before it gets to the store. Check out prAna's blog to read about how many of their clothes are now packaging-free, and watch a cool video. Then contact your favorite clothing company and challenge them to do the same.[116]

To find many more sources of natural and organic fiber clothing, check out the Clothing section of the Organic Consumers Association's Buying Guide.

And to learn everything you ever wanted to know about sustainable fabrics (including the truth about bamboo and other rayon fabrics [they're not as eco-friendly as you might think], how to keep natural fibers from wrinkling, why we should steer clear of nano textiles, how to find ethical wool, and much, much more) visit OrganicClothing.blogs.com.

Toys & Kids Stuff

While I don't have kids, I had a lot of fun looking up plastic-free toy options. At least while kids are young, there doesn't seem to be any reason to buy them plastic toys—including polyester plush toys which are actually plastic—because there are quite a few companies producing adorable toys from wood and natural fibers. Here are a few I found:

Stuffed Animals and Dolls Barefoot (Barefootceylon.com), Ecoleeko, The Earth Friends, and Global Green Pals are some really cute dolls and stuffed animals. Cate and Levi's makes adorable plastic-free animal puppets, purses, backpacks, hats, mittens, slippers, and scarves all from recycled sweaters and local wool. Owner Josh Title says, "My belief is that there's enough material in existence in the world that we could probably freeze all new production effective immediately and just get more creative with what's already out there."

Wooden Toys Maple Landmark, Heirloom Wooden Toys, Plan Toys, and Tree Hopper Toys are good choices.

Natural Teethers Little Alouette and Ringley Natural Teethers are made from unfinished natural wood. Zoe B. Organic offers teethers made from natural rubber instead of plastic.

Beach Toys Look for little metal buckets and shovels instead of plastic. Toys taken to the beach can easily wash into the ocean. If you do want a plastic option, consider the beach toys from Zoe B. Organic, which are made from plant-based PHA plastic, which (as I mentioned in Chapter 5) is the only plastic proven to biodegrade in sea water.

"My 4-year-old, Hannah, loves the toy catalogues. Sometimes I try to recycle them before she can see them, but she has this sixth sense that foils my efforts every time. If a major toy company or toy store has distributed promotional materials to our area, she can sniff them out. And as soon as she does, the asking starts. '*Mama! Mama! Mama! I want this one, and this one! And, of course, this one! Mama! Did you hear me? I have to tell you something! I want this one!*'

"Sometimes a friend or relative will buy her the plastic doll she wants. When the doll arrives in my home, I don't throw it out. But I also have the choice to not be the one who buys the doll in the first place, and I exercise that option. I believe that in doing so, I am making a difference. I am supporting an ethic that I care about, I am working to protect the planet, and I am communicating a message to my children about consumerism. Because one doll might not be all that bad, but millions of dolls purchased by millions of people because 'just one can't hurt' add up really fast.

"Cheap plastic toys are easy to find, and they are marketed aggressively to both us and our children. But we don't have to buy into that message. And we aren't ruining our children if we don't. We are protecting their childhoods, and teaching them what really matters." —Amber Strocel, www.strocel.com, Vancouver, Canada

Plastic-free cat toys

Online Children's Stores Bella Luna Toys, Hazelnut Kids, Wild Dill, and Mighty Nest are great resources.

Pet Toys

While there are a few companies making pet toys out of recycled plastic, I avoid those products because I don't want my kitties chewing on any kind of plastic, recycled or otherwise. Pam Wheelock, the owner of Purrfect Play agrees. Purrfect Play makes cat and dog toys from organic cotton, wool, and hemp. At our local pet shop, we've also found all natural wool, coconut, and feather toys, as well as cardboard and Pioneer Pet natural sisal cat scratchers and climbers as opposed to ones made with synthetic carpet. But our kitties' favorite toys are simply used wine corks, empty boxes, and Michael's toes when he comes home from the swimming pool. (Cats are crazy.)

Household Items

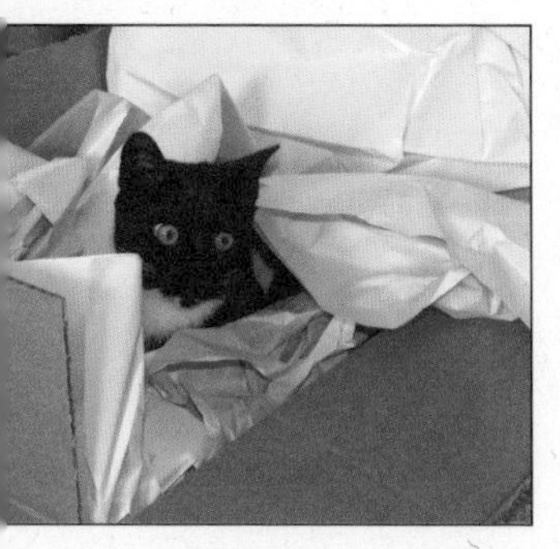

Cats prefer empty boxes to expensive cat condos anyway.

Bath Mats and Shower Curtains How do you keep from slipping in the shower and spraying water everywhere without plastic? Vermont Country Store carries natural rubber tub mats and 100 percent cotton shower curtains. Check Etsy for handmade cloth shower curtains made from organic cotton or hemp. Or make your own shower curtain out of an old sheet or table cloth, using your old shower curtain as a model. Cloth shower curtains do need to be washed regularly to keep mildew at bay.

"I cut worn towels into 1 inch strips, tie three strips together in a knot in one end, and braid them. I finish it off by tying a second knot at the end of the braid. This toy is cheap, plastic-free, and stands up to a dog named Chopper." —Lisa Nelsen-Woods, www.condo-blues.blogspot.com, Columbus, Ohio.

If you prefer a plastic tub mat or shower curtain, look for PVC-free options: tub mats made from TPE or shower curtains made from polyester or ethylene vinyl acetate (EVA). These shower curtains do not give off that "new car smell" indicative of the toxic chemicals added to PVC.

Natural Yoga Mats Yoga is a healthy practice, but not when it's done on toxic PVC yoga mats. Here are a few plastic-free alternatives:

- **Rawganique** Hemp yoga mat. Use alone or as a topper for a stickier mat.
- **Jade** Natural rubber yoga mats.
- **Eco Yoga Mat** Jute and natural rubber.
- **prAna Indigena Natural yoga mat** Natural rubber.

If you're allergic to natural rubber or don't like the smell of it, opt for PVC-free yoga mats made from thermoplastic elastomer (TPE). It's a less toxic plastic material.

Furniture & Mattresses Whether buying new or secondhand, try to avoid upholstered furniture and mattresses made with polyurethane foam. Furniture with a label stating it meets the California furniture flammability standard will most likely contain toxic flame retardants (See explanation in chapter 1.) Some other plastics to watch out for are Teflon stain resistant treatments on upholstery and PVC mattress covers.

When buying new furniture, consider untreated sofas and chairs with cushions made from natural latex or other plant-based foams and organic wool and cotton batting. A few furniture companies to check out are GreenSofas.com, Eco-terric, EcoBalanza, and Ekla Home. The least expensive of the bunch is GreenSofas.com, but even their couches are high compared to what you'll find at most conventional furniture stores. A 100 percent organic cotton futon sofa is a less pricey option. We've been sitting on an all cotton futon sofa for years, and while it might not be as soft as a foam couch, it works fine for us. No matter what you buy, check to see if it's been treated with PBDE flame retardants.

For less-toxic, less-plastic mattresses, check out OMI (Organic Mattresses Inc), Naturepedic, Earthsake, or MyGreenMattress.com. The Futon Shop offers organic futon options. Ask if the mattresses contain flame retardants. Mattresses that contain wool are naturally flame resistant and less likely to need flame retardant protection. Jennifer Taggart, author of *The Smart Mama's Green Guide* (Center Street), recommends letting conventional mattresses off-gas outside for a minimum of three days if possible before bringing them into the house.

To find more sources of natural and organic furniture and mattresses, check out the Home section of the Organic Consumers Association's Buying Guide.

Music, Videos, and Software—Just Download It! I can't remember the last audio CD I bought. Nowadays, I purchase mp3s from iTunes or other electronic music services. I stream movies and television shows from Netflix and other sites. And whenever possible, I download software and software upgrades instead of purchasing programs on CD-Roms, which will just end up as e-waste as soon as the next upgrade is released. As I mentioned before, CDs and DVDs are made from polycarbonate plastic, which contains BPA. And while most of us are not going around licking our CDs, I think it's important to reduce the demand for this chemical in the first place.

Electronics If you decide to buy brand new electronics rather than secondhand, there are several guides to help you choose the greenest models with the fewest toxic chemicals, highest recycled content, and least waste from the most responsible companies. Visit the Electronic TakeBack Coalition's "Tools for Purchasers" page for links to green electronics buying guides.

And look for electronics and components with the least plastic parts and packaging. One of my favorite finds are thinksound "ear bud"-type headphones, which are made from wood, metal, and silicone, have PVC-free cords, and are packaged in cardboard with almost no plastic.

Building and Remodeling Materials

Once you start noticing all the plastic around you, you can't help noticing how much plastic is incorporated into your home and the buildings you frequent on a daily basis. Our wall-to-wall carpet is a daily reminder that plastic is everywhere. While we can't all rip our homes apart and start from scratch, there are a few things we can keep in mind when planning any remodeling.

Rethink Carpet. Carpet is nice and warm, but personally, when I think about it for too long, I get grossed out. Carpet is fabric that we walk on every day, tracking in dirt and chemicals from the outside even if we religiously remove our shoes when entering the house. Carpet is much harder to clean than hard floors and it's often made from synthetic fibers with PVC backings and toxic adhesives. If you have a choice, skip the carpet. Choose sustainable FSC-certified wood flooring, bamboo, cork, or natural linoleum with non-toxic finishes.

"If doing a remodel or addition, use as many re-used fixtures as you can. New ones come wrapped in plastic, foam, etc." —Gene Anderson, www.diyinsanity.blogspot.com, Oakland, California

Choose Plastic-Free Blinds and Window Coverings. Many older mini blinds are made of PVC and contain lead.

Look Into Natural Fiber Insulation. Despite its name, fiberglass is plastic and may contain toxic flame retardants. Our friends Nancy and David insulated their crawl space with Bonded Logic's recycled denim insulation instead. Or, look for natural insulation made from recycled newspapers, hemp fibers, or magnesium oxide. Beware "bio-based" spray foams, as they also contain petroleum-based chemicals.

Say No to Vinyl Siding. Watch the movie *Blue Vinyl* for an entertaining yet eye-opening look at the toxic world of PVC, as filmmaker Judith Helfand attempts to convince her parents to replace the vinyl siding on their house.

Home insulation made from recycled jeans

Look For Salvaged and Reclaimed Materials. Check out these sites for more information about plastic-free and less toxic building alternatives:

- **U.S. EPA's Green Building page**
- **U.S. Green Building Council's Green Home Guide**
- **Seattle Green Home Remodeling Guides**
- **Green-Talk** Blogger and green building consultant Anna Hackman reviews home products and writes about her personal successes and failures remodeling her own home and living sustainably. She's funny, too.

Office Supplies and Miscellaneous Products

Check out these ideas for even more deplasticking solutions:

Avoid Disposable Plastic Pens. These days, I write mostly with FSC-certified pencils and carry a little metal pencil sharpener with me wherever I go. I invested in a Lamy refillable fountain pen with a converter cartridge, which allows me to fill the pen from a glass bottle of ink instead of a disposable pen cartridge. Using a fountain pen is not the most convenient way to write, especially on the go, which is why I reserve it for signing occasional checks or business forms.

Refuse Plastic-Covered and Coated Notebooks and Binders. I buy my binders and little notebooks from GuidedProducts.com, which sells office products made from 100 percent postconsumer recycled paper without any plastic or plastic packaging.

Get Off Paper Mailing Lists. Reduce paper consumption as well as plastic envelope windows. Many people assume that because envelopes are recyclable, the windows are made from a biodegradable material. In a few cases, windows may be made from PLA or glassine, a paper product. But more often these days, envelope windows are made from plastic, and it's not always easy to tell the difference. I do all my banking and bill paying online and have used services like Catalog Choice (also known as TrustedID Mail Preference Service) to get off mailing lists.

Try Making Your Own Paste or Glue. You can make a simple flour and water paste from ½ cup (60 g) flour, 3 cups (700 mL) boiling water, and ¾ (180 mL) cup cold water. Slowly pour cold water into flour and stir to make a paste. Pour paste into the boiling water, stirring constantly. Cook for 5 minutes or until the paste is thick and smooth. When cool, pour into an old squeeze-top glue bottle and store in the refrigerator. There are many other paste and glue recipes on the Internet.

Refill Ink Cartridges. While HP takes back its printer ink cartridges to be recycled, refilling is less expensive and has a lower environmental impact. Services like Cartridge World sell refilled cartridges and guarantee their performance. If you do a ton of printing, you could also consider investing in a Silo Ink system, which saves even more ink and plastic. The Silo system itself is made from plastic, but saves more plastic in the long run. The greenest option, of course, is not to print if you don't have to.

Consider Refillable White Board Markers. AusPen makes a refillable marker system with reusable pens made from recycled plastic and xylene-free ink. Sure, the ink comes in plastic bottles, but each bottle holds much more ink than a regular disposable plastic marker.

Say No to Plastic Lighters. Disposable plastic lighters are a common site in photos of dead albatross chicks on Midway Island. If you need to light a fire, opt for matches or a refillable metal lighter.

Additional Shopping Resources

Here are a few more resources for finding greener products with less plastic. Be sure to request no plastic packaging when ordering.

- **Green America's Shop Green Pages**
- **BuyGreen**
- **Lehman's**
- **Franklin Goose**
- **Kaufmann Mercantile**

No More Plastic Shipping Materials

How would you feel if you opened your front door to find a huge box from your favorite fudge sauce company sitting on your front steps, and you were pretty sure you hadn't ordered it? Sweet, right? But wait, there's more. Imagine you open the box to discover that while it does contain twenty glass jars of the best fudge sauce ever invented, it's mostly full of Styrofoam peanuts and bubble wrap. Now how much would you pay?

Turns out, the fudge was a surprise from my dad, who had no idea how it would be packaged. I was thrilled to accept the gift but not so happy about the plastic. So, realizing that the fudge sauce company was located in San Francisco, I decided to call and ask if I could return the Styrofoam to them to reuse. Here's how the telephone call went:

Ring

Elderly woman's voice: Hello?

Me: Um . . . is this the Fudge Is My Life Company?

She: Oh yes, sorry. I thought you were going to be someone else.

Me: Oh. I'm a customer. Have I reached your home?

She: Yes, in fact my office is in my home. How can I help you?

Me: Is this actually Lillian Maremont [the founder of the company and creator of the original recipe back in 1963]?

She: Yes, that's me! [laughing]

Me: Well! I just have to tell you how awesome your fudge sauce is. It's the best I've ever had. I love it so much. I just received a case in the mail from my dad.

She: Oh, do you live in Oakland? I remember your order. I processed it myself.

Me: [Remembering the original purpose of my call . . .] I do love the fudge sauce, but it came packaged with all these Styrofoam peanuts and bubble wrap, and for environmental reasons I don't want to throw it away, so I was wondering if I could bring it back to you to reuse.

She: Oh, well, I guess you could. But I live out in The Avenues in San Francisco . . . kind of far out for you. Why don't you take it to the warehouse company we use in Emeryville? [She gives me the name and address of the warehouse, as well as directions for how to get there.]

Me: Okay, I'll take it to them. Just wondering . . . have you considered using biodegradable packaging rather than Styrofoam?

She: We would love to, and we have tried the corn-based peanuts. But they're just too expensive for us right now. And I'm not sure any other type would protect the glass jars enough. I just really want to make sure that my fudge arrives in perfect condition for my customers to enjoy.

Me: Well, I hope the price comes down for you so you can switch away from the Styrofoam, and I'll take this material over to the warehouse.

And then I gushed some more about the fudge sauce, and she said I made her day. We both hung up happy. A few minutes later, I strapped that huge box of plastic onto the back of my bike (Oh yes I did!) and pedaled across town to the warehouse, where the receptionist very graciously accepted it from me and promised to reuse it. Sure, it was extra work on my part. I could have just taken it to the UPS shipping center a few blocks from my house, and the employees there would have gladly reused it as well, but returning packaging to the companies that sent it lets them know that their customers would prefer they come up with more eco-friendly alternatives. How else will they get the message?

It's surprising to me that so many companies producing eco-friendly, otherwise plastic-free products still package them in plastic to ship them out. As I've suggested several times in this chapter, it's always a good idea to request no plastic packaging in advance when ordering online—even from green companies! Sometimes companies will ask for suggestions of what to use instead. Here are a few examples:

- **Yesterday's News Padded Mailers** from the Sealed Air company are made from recycled newspaper fiber.
- **Jiffy Padded Mailers**, also from Sealed Air, are made with recycled paper pulp padding rather than plastic bubble padding. Note: Sealed Air makes both paper and plastic padded mailers.
- **Jet-Cor Rigid Corrugated Mailers** from iVex Protective Packaging are 100 percent cardboard.
- **Biodegradable Cellulose Packing Tape** is made from wood pulp with a natural rubber adhesive. Look for it at EcoEnclose.com or Life Without Plastic.
- **Kraft Paper Tape** is available in many office supply stores.
- **Molded Paper Pulp Packaging** from UFP Technologies is made from recycled paper.
- **Ecocradle Molded Packing Material** from Ecovative is made from mushrooms.
- **PaperNuts** are pellets made from recycled paper.
- **Geami GreenWrap** is protective wrap made of paper that takes the place of bubble wrap.
- **ExpandOS** are little expanding paper pyramids that replace Styrofoam peanuts and other packaging materials.
- **Reused Packaging Material**, of course, is the option with the least environmental impact. At home, I don't buy any of these products but just use old newspaper and boxes and envelopes I've received in the mail.

Reduce, Reduce, Reduce

The biggest lesson I have learned since I started my plastic-free journey is the joy of less. Of buying less stuff and making do with the stuff I already have. While acquiring new things can be momentarily satisfying, I'm ultimately left with a bunch of meaningless stuff—stuff to organize, to maintain, to fix when it breaks, to trip over when it's left lying around, and to deal with at the end of its life. I feel a real sense of freedom and relief when I come home from a trip with only a few digital photos and memories of the experiences I had rather than the time I spent shopping for souvenirs. I love going through my closets and weeding out things I no longer need, offering them to friends or people in the neighborhood via Freecycle or donating them to Goodwill so someone else can make use of them. The worst part of having a lot of stuff is figuring out what to do with the crap that no one wants, usually broken plastic items that can't be fixed or even recycled and have no more apparent usefulness left in them. Learning to live simply with a few, good quality, multifunctional items and resisting the urge to buy more is challenging in our acquisitive, consumerist society, but ultimately satisfying.

I've talked about saying no to single-use disposables and single-serving sizes, but there is one more "single" I've learned to refuse: single-purpose items. Across America, kitchen drawers are stuffed full of gadgets that serve one purpose and one purpose only, like saving a single piece of fruit or vegetable, for example. There's the plastic tomato-shaped "tomato saver," the avocado-shaped "avocado-saver," or the banana saver shaped like an orange. I mean, banana. Then there's the plastic lettuce knife, which is necessary since lettuce can't simply be torn with the hands, or the Butter Boy, a plastic gadget meant solely for buttering corn. Erin Doland from the website Unclutterer.com writes a weekly segment called "Unitasker Wednesday," in which she highlights the worst of single-purpose gadgets that not only waste resources but simply clutter up our lives. Among the gems she has found are the mayo knife spreader, the potato chip finger, the watermelon cooler, and my personal favorite, the Krustbuster: a hunk of plastic created specifically for taking the crust off slices of bread. *Are. You. Kidding. Me?* If you must waste perfectly good food, at least do it with a ***knife!*** I could waste hours ogling the stuff on this site and laughing at its ridiculousness if it weren't for the troubling fact that these things wouldn't exist if people weren't buying them.

> "Just say no to freebies at expos and conferences. You'll be amazed at how much less plastic waste (and otherwise) comes into your home, and how much lighter your luggage will be!" —Debra Baida, personal organizer, www.liberatedspaces.com, San Francisco, California

Of course, not all single-purpose items are useless. A coffee maker makes coffee, and that's all it does. I'm okay with that. But there are so many other things that seem like fun in the moment and end up cluttering our lives. The key is to take a moment before we click the order button or pay the cashier to ask ourselves not only "Do I really need this?" (Let's face it, needing something is not the reason many of us shop.) but, "Will this really make me happy?" Will having a white iPhone bring me twice as much joy as the black iPhone I already have? Will upgrading to a camera with a couple more megapixels bring me closer to the people I want to take pictures of? Will adding one more stainless steel lunch container or organic cotton bag or reusable bottle to my already massive collection of eco stuff make me greener than the guy next door? If this thing that I am tempted to buy is not going to bring me real happiness, then what will? What is missing in my life that I'm trying to fill with more stuff?

Here are a few resources and reasons for living with less.

- **Center for a New American Dream** emphasizes simplifying your life, rejecting consumerism, and focusing on what matters.
- **No Impact Project** Inspired by the book and film of the same name, the project invites people to experience no-impact living for one week.

- **The Story of Stuff: How Our Obsession with Stuff Is Trashing the Planet, Our Communities, and Our Health—And a Vision for Change,** by Annie Leonard. Free Press, 2010.
- **What Would Jesus Buy?** This documentary film is a comedic yet heartfelt crusade against America's overconsumption and commercialization of Christmas.

QUICK LIST: QUESTIONS TO ASK

1. Do I already have something similar stashed away that I could use instead?
2. Can I repair what I already have?
3. Can I find it secondhand?
4. Can I borrow or rent it?
5. Can I download it instead of buying a CD or DVD?
6. Can I build, sew, knit it myself?
7. Can I check Etsy.com or other resources to find someone to make it for me?
8. Can I find a refurbished version?
9. Can I find a plastic-free or less-plastic version?
10. Can I find it made of recycled materials?
11. Can I buy the least toxic/most ethical brand?
12. Can I find it without plastic packaging?
13. What will happen to it at the end of its life? Will the manufacturer take it back?
14. Will having this item truly make me happier?

SIMPLE LIVING HERO:
Erika Barcott

In 2006, Erika Barcott moved from her 1,200-square-foot apartment in Seattle to a 400-square-foot cabin in the country. I met Erika a few years later after posting a plea for help on my blog. Even though I had stopped buying new plastic, I still had an epic pile of crap in my home office that was threatening to capsize on an unsuspecting kitty one day. I needed help organizing and asked my blog readers for ideas. Erika, who had had to learn a lot about the subject after moving into such a small space, chimed in with great suggestions and even lent me a book on de-cluttering, but the best advice she gave me was about learning to live with less stuff to begin with.

See, Erika's Seattle apartment had been full of stuff—stuff she'd bought impulsively without having a clear plan for how she would use it. When she decided to move into a space one-third the size, she really had no idea how much stuff she actually owned and figured she would just rent a storage space until she figured what to do with it. Fortunately, Erika didn't opt for that quick fix and instead, crammed all her boxes into the tiny cabin. After living in such crowded conditions for six months and unable to get any of it organized, Erika realized she would have to do some serious downsizing. And that proposition brought up a lot of tough emotions—Erika was very attached to her things. So she started reading books on decluttering and looking at decluttering websites. Finally, she came up with a plan.

Erika examined her lifestyle to assess what activities gave her the most joy and satisfaction and set some priorities. Two of her passions are reading and knitting (Erika blogs at RedShirtKnitting.com), so she knew she had to allocate a generous amount of space for her yarn stash and a reasonable number of books. Those priorities meant that she wouldn't have room for a dining table. So, instead of inviting friends over for dinner, she goes out to meet them once a week, which is good for her since she works from home and otherwise might not get out much. About 20 percent of her boxes contained broken stuff that she thought she might fix "some day." That was the first stuff to go. Next, she measured her bookshelves and decided that she would only keep as many books as would fit on the shelves. She carted off several boxes of books to used bookstores.

It took Erika a while to reduce her possessions to an amount that would comfortably fit in her space because she wanted to dispose of her things responsibly. She says, "In our society, we get a lot of messages about how to buy but not about what to do with it when we don't want it any longer." Slowly, she found homes for most of her possessions, and now that her stuff is down to a "space-appropriate" amount, she feels at home. I asked Erika how she felt while eliminating so many things from her life. She said that she went through continuous emotional cycles—fear and anxiety as she made decisions about what had to go, followed by giddy highs when she finally dropped things off at the thrift store. But in the end, downsizing has had many more benefits than drawbacks. She eats a lot better because she doesn't have a freezer for convenience meals and doesn't have space to store a lot of packaged foods. (That means she also generates less waste.) She never has to face a sink full of dirty dishes because she doesn't have many dishes in the first place. She enjoys the few decorative items she kept because they are out where she can see them instead of tucked away in boxes. And because her rent is so much lower, she doesn't have to work as much and spends a lot more time enjoying her life.

We don't all have to give up our homes and move into a small space to experience the freedom of living simply. What if those of us blessed with more room than we need lived *as if* we weren't? How would our habits of buying and spending change? How much extra stuff would we allow to consume us? As Erika says, "Doing without is not about deprivation but about consciously examining our priorities and values and mindfully choosing those things that really make us happy."

Action Items Checklist

(Choose the steps that feel right to you. Then, as an experiment, challenge yourself to do one thing that feels a little more difficult. Only you know what that one thing is.)

- ☐ Gather up toys and other children's items made with PVC for toxic waste facility.
- ☐ Look up PVC-free alternatives on CEHJ's Guide to PVC-free school supplies.
- ☐ Contact local schools and hospitals about the hazards of PVC.
- ☐ Set priorities for phasing out plastic kitchenware.
- ☐ Obtain plastic-free replacements for kitchenware.
- ☐ Gather up existing plastic foodware and repurpose or donate.
- ☐ Research local sources for borrowing, renting, and buying used.
- ☐ Analyze your wish list and figure out what can be obtained second-hand or borrowed.
- ☐ Mine your junk pile for treasures that can be repaired and put to use rather than replaced.
- ☐ Remember to request zero plastic packaging when placing orders.
- ☐ Return unwanted packaging to vendors.

Chapter 10: Plastic-Free Living Around the World

When I started my plastic-free journey back in 2007, there were only three other bloggers I knew of doing something similar: Envirowoman in Canada (whose blog no longer exists), Martin Higgins in Malta (whose blog, Plasticless, is archived as a resource online although he no longer updates it), and Kate (a.k.a. Polythene Pam) in the UK, who you'll read about later in this chapter. But as far as I knew, I was the only person in the United States attempting this lifestyle, and although I was slowly gaining blog readers, I still felt pretty alone. Maybe even a little crazy at times. I mean, only four of us on the planet?

Eight years later, the plastic-free movement has gained a lot of traction. All over the world you'll find bloggers and activists reporting on their experiences reducing plastic where they live. And while it would feel satisfying to attribute the growth of the movement to the impact of my book and blog, doing so would be nothing short of delusional. People are coming to it on their own after getting fed up with all the crappy plastic products and packaging they are exposed to every day, and are seeking out other like-minded folks for support. And I'm just happy to be one small part of this tidal wave of change.

Here are just a few of the plastic-free bloggers and activists around the world. Visit MyPlasticFreeLife.com/blog-roll/ for an up-to-date list. Hopefully, these individuals will inspire you to take on your own plastic-free living challenge and maybe even blog about it, no matter where you live.

AUSTRALIA

If any geographical group has had the most success in getting a lot of people to try living plastic-free, it's probably the Australians. Well, the Western Australians to be exact, and one woman in particular.

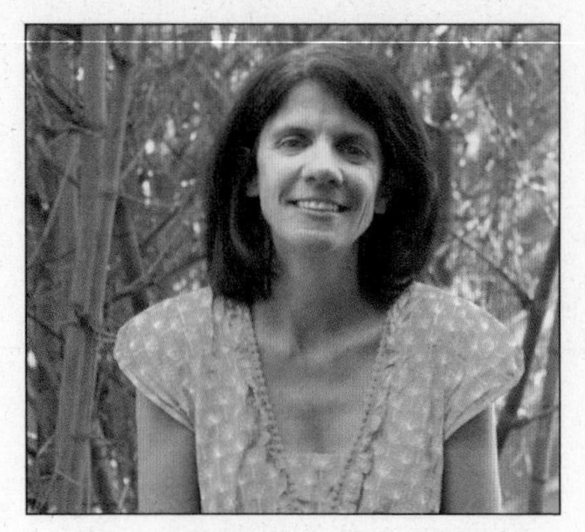

Rebecca Prince-Ruiz, founder of Plastic Free July in Western Australia

Name: Rebecca Prince-Ruiz
Website Name: Plastic Free July
Website URL: www.plasticfreejuly.org

Working for the Western Metropolitan Regional Council, Rebecca Prince-Ruiz was part of a team running an adult education course to help people understand the proper way to recycle. As part of the course, they would take participants on "the journey of their waste," including visiting the local recycling facility. Rebecca says she'd seen a landfill before, but at the recycling center, she was "completely blown away."

> I couldn't believe the sheer volume of how much recycling we produce and the complex and energy intensive process that goes into transporting, sorting, retransporting materials for recycling . . . and this last stage often happens across the other side of Australia or overseas. In an instant recycling went from a "feel-good green action" to a new mindset where every time I went to put something in my recycling bin I felt filled with remorse and wondered how I could have purchased differently. Don't get me wrong, I think recycling is a good thing, but it's a choice of last resort—refusing, reducing, and reusing are much more sustainable options.

So she turned up to work one day in 2011 and said, "Hey, let's go plastic-free next month." They put the word out in their newsletter and got forty households to participate. At the end of the month, several participants met together and shared what they had collected in their "Dilemma Bags"—the plastics they hadn't been able to avoid. They gave each other support and shared stories of successes and failures, recipes, and ideas, and recommended shops. Several people went on to found other plastic-related projects. The next year, in 2012, they created a Facebook page, showed the film *Bag It* in a local school, and organized a workshop with demonstrations on things like making your own yogurt and toothpaste and making reusable produce bags from old curtains. More and more people joined in.

In 2013, they built a website, Plasticfreejuly.org, and the campaign spread across the world. People could sign up for the challenge for a day, week, or the whole month and become part of a community of inspiration and commiseration. That year, only 47 percent of participants were from Australia. In 2014, over 14,000 people from sixty-nine countries participated and Rebecca hopes those numbers will continue to grow. In fact, I myself was so inspired by Plastic Free July that year, I worked with the Ecology Center in Berkeley, CA, to start our own local Plastic Free July group, with events that included a movie night, presentation, show your plastic challenge, and plastic-free potluck. I'd love to continue doing it in future years.

I asked Rebecca to tell me a little about her challenges, successes, and discoveries trying to live plastic-free in a suburb of Perth, Western Australia. She wrote:

> Some of my biggest challenges revolve around my kids. Overall they are great and really supportive yet there are so many things they like that we can no longer buy. For instance they like Mexican food—all the corn chips come in plastic coated foil as do burritos and sour cream. It's a fine line between my values and at times compromising in order not to put them off. We now make all our own bread, biscuits, cakes, yoghurt, and preserve a lot, but the thing I miss most is teabags! This year I learnt that most teabags (even the ones that look like paper) are about 30 percent plastic!
>
> I'm lucky to live in Fremantle where there are already a few shops selling bulk produce. My favorite plastic-free discovery is my little pouch of netting "produce bags" I got from Onya Australia which is a local company. They are a great lightweight addition to my handbag, which I use daily to purchase fruit and vegetables, nuts, grains, coffee, etc. Through our Plastic Free July campaign we

have a list of local places on our website and with the help of participants can continually update. One of the best things about being plastic-free is that it means you avoid the large corporately owned supermarkets and support small local businesses—we take our own containers and bags to the local baker, greengrocer, butcher, fishmonger, farmers market, and delicatessen. Get organised, be prepared, tell shopkeepers why you are trying to avoid plastic and don't be shy!

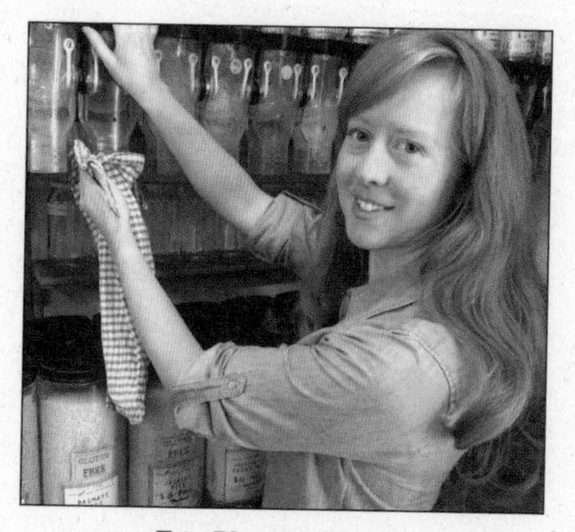

Erin Rhoads, The Rogue Ginger, Melbourne, Australia

Name: Erin Rhoads
Website Name: The Rogue Ginger
Website URL: www.therogueginger.com

In 2013, across the country in Melbourne, redheaded Erin Rhoads had just watched the film *The Clean Bin Project*, and while the film didn't specifically deal with plastic, there were parts that highlighted the effects of plastic pollution. Erin didn't like what she saw and, in her words, "I was hit with the realisation that I was part of the problem. Instead of dwelling on it I decided to become part of the solution. The moment that movie ended I began researching what I can do to make better choices, and that is when I discovered Plastic Free July."

So she started a wonderful blog called "The Rogue Ginger," where she documents her strategies for plastic-free living in her area, complete with beautiful photographs. She and her partner now shop at bulk food stores and use their own packaging to collect items like food and cleaning products. They stopped getting takeaway food. And Erin makes all of their personal products and makeup. In fact, she created a recipe for homemade mascara with only organic soap, activated charcoal, purified water, and almond oil. The instructions are on her blog.

Erin says that Melbourne and all large cities in Australia have a growing number of bulk food shops where you can bring your own bags and containers for packaging-free food. Local shops in the Melbourne area include Friends of the Earth Food Co-op in Collingwood, Organic Wholefoods in Brunswick East, and The Source Bulk Foods in Prahran. Elgar Dairy offers products in returnable glass bottles. Queen Victoria Markets and Flemington Farmers Markets are great for plastic-free vegetables, fruit, eggs, fresh pasta, and bread. And Perfect Potion allows shoppers to take their own containers for clay, cocoa butter, and beeswax, a great option for those wanting to make their own beauty products.

Recently, Erin discovered a resource for people living in New South Wales. No Plastic Fruits and Vegetables is Australia's first plastic-free grocery store. Unlike most natural food stores that sell both bulk products as well as products packaged in plastic, Erin says that this store's goal is to actively encourage shoppers to understand that plastic is not necessary. You can read Erin's interview with store owner Alex Grant on her blog.

Erin says that the main thing she still struggles with are [BPA-coated] receipts and plastic medication containers. (Me too, Erin, me too.) But for the most part, they have eliminated disposable plastic from their lives and have discovered some unexpected benefits:

> Going out for dinner or sitting down at a café to enjoy a hot chocolate has become so much more special. I no longer take food home as takeaway or rush about with a drink to go. Going plastic-free has given me a permission slip to not rush but rather take time to sit and enjoy. I can't give into regular indulgences like I used to which in turn means I save money.

Erin plans to start encouraging local businesses not to hand out plastic straws with drinks. And she's also working with her sister to create Plastic Free Living Starter Kits so that people in her area can make the step to reduce their plastic with more ease than she had. Her most important piece of advice to anyone trying to go plastic-free is to not be afraid to contact local bloggers and ask questions if you're stuck on a particular item. "I can tell you, we are all happy to help another human being reduce their plastic waste."

Name: Lindsay Miles
Website Name: Treading My Own Path
Website URL: treadingmyownpath.com

Lindsay was inspired by Plastic Free July in 2012. She and her husband committed to continuing their plastic-free lives after July was over, and at the closing ceremony for Plastic Free July 2013, they were asked to speak. Lindsay provides helpful information on plastic-free, waste-free, sustainable living on her site.

NEW ZEALAND

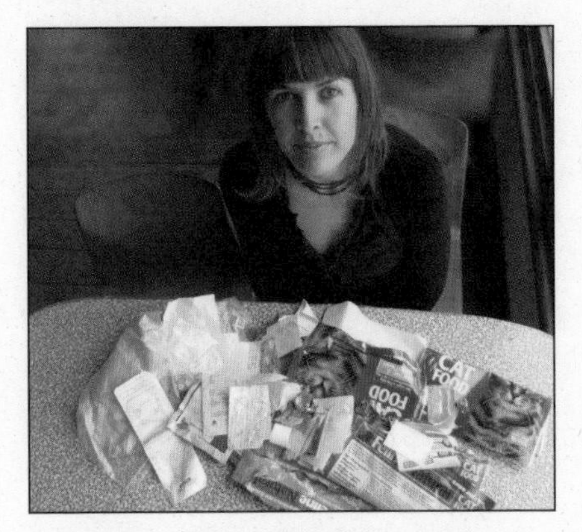

Merren Tait with her one year's worth of collected plastic trash, Raglan, New Zealand

Name: Merren Tait
Website: A Year without Buying Plastic
Website URL: 1yearnoplastic.blogspot.com

Just across the Tasman Sea, Merren Tate from Raglan, New Zealand, was also inspired by a film. In 2013, after watching the trailer for the documentary ***Midway: Message from the Gyre***, which exposes the tragic effect plastic pollution has on an albatross colony, she started researching projects designed to curb plastic waste and discovered Plastic Free July. Thinking it would be a challenge that people in her community would be receptive to, she wanted to do something a little dramatic to promote it and give it a high profile. So she decided to challenge herself to not buy any plastic at all for a whole year. It worked; Plastic Free July has been a successful annual initiative in Raglan.

Merren chronicles her efforts on her blog, where she provides an alphabetical shopping guide. Here are a few local resources she recommends:

Bin Inn is a nationwide New Zealand grocery store chain specializing in natural and gluten-free whole foods that you can purchase from bulk bins in your own container. Bin Inn won the 2012 Unpackit NZ Packaging Award for minimizing packaging waste. If there's no Bin Inn near you, Merren says most supermarkets and health food stores do have some bulk bins, although their selections can be limited.

Go Bamboo is a fledgling company making bamboo products. Currently they produce toothbrushes, cotton buds, and clothes pegs but are developing more products.

Greencane makes toilet paper, paper towels, and toilet tissue out of sugarcane bi-product. They use compostable packaging, are stocked in New World supermarkets, and offer free delivery to your home.

Ecostore produces toiletries like soap, shampoo, and moisturizers, as well as laundry and baby care products. Shop online or visit one of their stores where they have bulk buying options. The have just developed bio-plastic packaging made from sugar cane that is 100 percent recyclable and captures carbon in its production. Most supermarkets stock their products.

Tina Ngata, the Non-Plastic Māori, New Zealand

Name: Tina Ngata
Website Name: The Non-Plastic Māori
Website URL: thenonplasticmaori.wordpress.com

Tina Ngata is a Māori woman living in Tūranganui A Kiwa (Gisborne), a small city on the East Coast of the North Island of Aotearoa/New Zealand. She began her plastic-free journey when several different events converged to show her the connection between working to reduce plastic pollution and the ancestral practices and philosophy of her Māori heritage.

According to Tina, her city has twice the national average of plastic in its waste stream, which she witnesses on her daily walks on the beach with her dogs. She says she'd always been aware of plastic trash on the beaches and felt responsible for picking up what she could. But the scope of the problem really hit home for her the year she followed the 2011–12 ***Te Mana o Te Moana*** journey—seven ocean-going twin-hulled ***waka*** (traditional ocean voyaging vessels) which circumnavigated the Pacific Ocean using ancient navigation techniques to raise awareness about ocean acidification and pollution. The crew blogged about their journey and the ocean trash they encountered along the way. Around that time, she came across another group, called "Para Kore," which aims to help all of their ancestral gathering spaces (marae) become waste-free. Their work resonated for her: waste-minimization as a means of reconnecting to ancestral practices.

Not long after that, she watched the short trailer videos for the same film Merren Tait had seen: ***Midway: Message from the Gyre***, and, like me in 2007 when I saw the photo of the dead albatross chick, she cried. A lot. Tina explains that the Māori call the albatross "Toroa" and revere it as an ancestor that keeps their voyagers company at sea and guards over them. They have an artistic pattern named "Roimata Toroa"—the tears of the albatross—and they adorn their most precious garments and prized possessions with feathers gifted from Toroa. So as she sat and watched Toroa die over and over again in that film and saw Toroa mothers unwittingly killing their own young with the waste of our human greed, she realized that she didn't want to contribute towards that waste anymore. She researched others who had taken plastic-free journeys and started her own on January 1, 2014.

Where Tina lives is geographically isolated from much of the North Island, which presents challenges in finding plastic-free options. She says their one bulk supplier doesn't have a particularly large range and there is quite some time between deliveries. Still, she couldn't imagine her journey without that one bulk supplier. The local farmers market isn't plastic free—most of the produce comes prepackaged in plastic bags—so she's had to learn to ask for what she wants, and most vendors are willing to comply. It also provides an opportunity to talk to them about what she's doing.

In addition to reducing plastic in her own life, Tina has reached out to try and encourage businesses to reduce plastic too. Early in 2014, she worked with three local cafés and an environmental group on a project called "Buy One Get One Tree." The cafés promoted plastic reduction and reported on all of the people who brought in their own coffee cups for a month, and at the end of the month they planted a native tree for each of them. They managed to save nearly 500 coffee cup lids from entering the waste stream. Tina is also working towards a plastic bag ban in her city and support for the initiative is growing. And she was invited by UNESCO to participate in the 2014 World Conference on Education for Sustainable Development in Japan to carry the message of indigenous rights and wisdom on sustainability. She says:

> This journey has, without a doubt, brought me closer to my ancestors. It's shaped how I see the world and how I see myself in it. By walking this path as best I can, and being open and frank about the challenges and successes, about my imperfect journey, and about how this relates to my culture, my identity, and hopes for the future, I have made an impact on those around me, and those who have heard my story, and that impact is growing every day. Kia Ora.

CAMBODIA

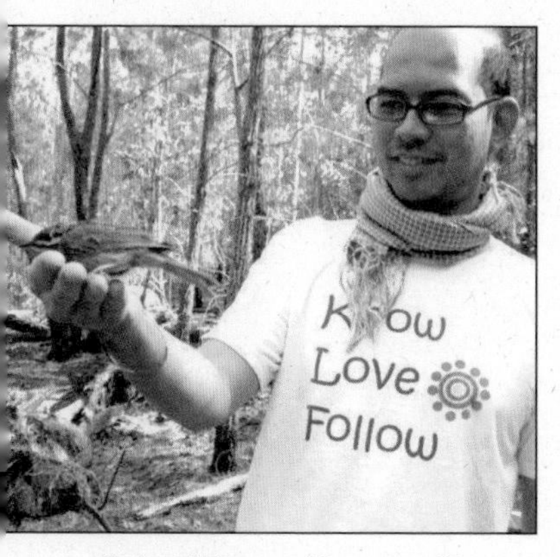

Gabriel Lamug-Nañawa, SJ, Plastic-Free Lent, Cambodia

Name: Gabriel Lamug-Nañawa, SJ
Website Name: As of the writing of this book, the website is under construction. Maybe it will be finished by the time you read this!

In early 2013, I received an email from Gabriel (Gabby), a Jesuit priest living and working twenty kilometers west of Phnom Penh, Cambodia, at the Banteay Prieb Center for the Disabled. He had written up a proposal for churches in his local area to promote Plastic-Free Lent. He himself had gone on a plastic fast for Lent the previous year and found the experience to be not only good for the planet, but an important part of his spiritual practice as well.

He wrote:

> Cambodia has had a tragic past and has only quite recently begun to open up to the modern world. Our cities are urbanizing very rapidly. But as people reach out and embrace modernity, a lot of other things such as disposable plastic is seeping through and is turning Cambodia's beautiful rustic landscape into a littered mess. Cambodia does not need these problems, and perhaps this coming Lenten season will be a good opportunity to pray and discern together how better still to serve this beautiful country.

Gabby's concern about plastic grew gradually with the increase in articles and awareness about the destructive effects of disposable plastic. He read an article and saw a feature on TV about the garbage patches in the ocean. He saw pictures of deceased birds with plastic in their stomachs. He visited the garbage dump in Cambodia and met the people who scavenge through it in search of recyclable materials. And he dived with majestic sea turtles in the Philippines and Australia and established "friendships" with them and then saw how they are potential victims to plastic bags that look like jellyfish.

During Gabby's plastic fast in 2012, the first items to go were instant noodles, which are very popular in Cambodia, followed by cookies, candies, and other snacks that come in throw-away wrappers. He began to use his own containers for things he bought and looked for food that was more traditionally cooked and packaged (like rice cakes wrapped in banana leaves.) Sometimes it meant giving up a tasty snack that he might have wanted but didn't actually need. And that was beneficial in forming his experience of Lent. I asked him what other challenges there are in reducing plastic where he lives and he said something I hadn't considered. He explained that poor people have "smaller portions of the world. They have smaller pieces of land (if any), smaller houses, using smaller amounts of soap, medicine, food. All of these smaller things come in smaller packages, most often in disposable plastics. It is not easy to be poor and to avoid plastics, especially in our context here in Cambodia." He focuses on encouraging people to refuse accepting thin disposable plastic bags that all sellers give away very freely even if you are buying a small thing like one bar of soap and refusing bottled water.

In fact, after giving each staff member a reusable bottle, they banned bottled water in the office. If a staff member purchases drinks in a disposable bottle, he/she is fined 1000 Cambodian riels (around 25 US cents). This amount goes into a can, their fund for a meal out for the whole team at the end of the year. (But if the plan succeeds, they won't be eating much during that meal.) Other steps they've taken have been to prohibit disposable plastic bottles during Jesuit and other meetings and to refuse to provide plastic writing pens. Finally, they are planning a study of disposable bottled water in Cambodia. The plan is to investigate four or five of the top local brands and send the water to a laboratory for water quality testing. They'll also study the success of recycling plastic bottles and what happens to those which are not recycled and are just burned by the roadside. The results of the study will be put into a PowerPoint presentation and shared with their students and communities that they work with.

To read more about Gabby's plan for Plastic-Free Lent, visit MyPlasticFreeLife.com and type "Giving Up Plastic for Lent" in the search bar.

TAIWAN

One cool and totally unexpected perk of my plastic-free adventure has been the opportunity to travel to different communities to give presentations on plastic-free living and learn what resources exist in those areas. But it wasn't until 2014 that I had an opportunity to travel overseas. In February, I received an invitation from a Taiwanese environmental organization called The Society of Wilderness to come and speak at a youth conference on marine plastic pollution at the National Museum of Marine Biology and Aquarium in late summer, all expenses paid. How could I say no?

That conference was groundbreaking. Unlike other youth conferences I've attended, in which kids listen to speakers and present information about projects they are working on, these Taiwanese kids had three days to organize an environmental action in the local community (working with kids they had just met!), carry out that action, and then create a multimedia presentation to show us, their mentors, what they had accomplished and what they had learned. The projects were fantastic. One group organized a flash mob, complete with original song and dance, to educate the public about the plight of hermit crabs that mistake plastic bottle caps for shells. Another group visited the vendor stalls at the night market and asked the sellers to post signs letting customers know that they would not hand out plastic bags unless specifically asked. Yet another group created a YouTube video about reusing plastic bags instead of taking new ones. There were groups with kids of various ages and levels of experience in public speaking, and they all gave their very best effort. I came home energized and inspired.

An addition to creating community actions, the participants were expected to live as plastic-free as possible during the conference. Food was served buffet style and everyone was expected to have their own eating utensils (chopsticks), reusable cup, and food container. Water and tea were served from big stainless steel dispensers. There were buckets for compost in addition to trash and recycling and loofah sponges in the bathroom for scrubbing off plates. The attention to detail was impressive.

Traditional market in Kaohsiung, Taiwan

Before leaving Taiwan, I wanted to see what other resources existed for shopping with less plastic packaging in Taiwan, so Jason Hu of the Society of Wilderness gave me a tour of the traditional markets in Kaohsiung. I saw bin after bin of loose rice, beans, dried mushrooms, and all sorts of other foods sold without packaging.

Before the trip, I was introduced to a Taiwanese blogger attempting to live as plastic-free as possible. Here is her story:

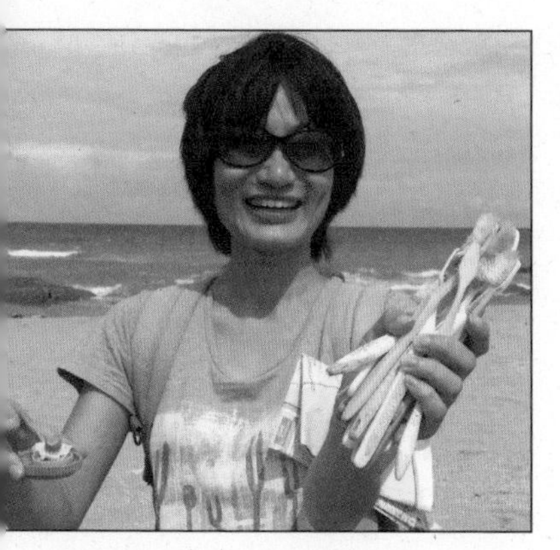

A-Hao, Simple Eco Life, Tainan, Taiwan, holds littered toothbrushes collected during a single beach cleanup.

Name: PING-SHAN, HUNG (A-Hao)
Website Name: Simple Eco Life
Website URL: simpleecolife365.blogspot.tw

A-Hao told me that she had been bored with modern life and having everything prepared by someone else. She wanted to create and have her own authentic experiences. It was then that she read the book ***No Impact Man*** and discovered "My Plastic-Free Life." It surprised her to find people actually trying to live plastic-free rather than just publishing theories, and she decided to try for herself. Her biggest challenge has been reducing food packaging. Despite the availability of foods from the traditional markets, most food comes heavily packaged in plastic—especially certified organic foods. And because of the humidity (which I can attest to!) things like flour, sugar, and salt are packaged in plastic to keep them fresh. So she joined a local food cooperative and is working on a food packaging campaign. She's trying to come up with other ways to package food in Taiwan, or at least to use safer plastic than PVC.

A-Hao suggests trying local farm markets. And she also recommends buying things together with friends. One reason so much food packaging is needed is that foods must sit on the shelf for a while waiting to be purchased. But if a group of people buy from farmers directly, they can request no plastic packaging and then have everyone pick up their purchases at the same time.

She also told me that in Taiwan it's often less expensive to eat restaurant take-out food than to cook at home, so there is a big culture of disposable restaurant packaging. She brings her own containers, but she suggests for people who are not comfortable doing that to have a seat. Choose a restaurant that serves food on reusable dishes for customers eating in.

For durable goods, there are hand-made markets, where you can find products made from wood, cotton, or metal. She herself opened her own shop in Tainan. Called Simple Eco Life, the shop carries stainless steel straws, bamboo toothbrushes, cloth pads, and other products. Customers can shop in person or online via her website. There's another shop in Taipei called Nature Miffy that sells lots of plastic-free products as well.

And for personal care, while she hasn't found a lot of plastic-free products, her suggestion is to stop using so many products in the first place. Her favorite plastic-free discovery is that often water is all you need for hair and skin cleansing. A-Hao says:

> We call it "poo-free" in Taiwan. It's an idea that clean water is all we need when we shampoo. Chemical shampoo removes the natural oils produced by the scalp and it causes the scalp to produce more oil to compensate. Washing our hair in a mild way will help the scalp become balanced again and also keep us away from harmful chemicals. After this, I also found some Japanese dermatologists suggesting that people wash their face without cleanser. Only warm water is enough (around 32–36°C). After face cleansing, moisturize with some coconut or olive oil. I've tried both ways for one year, and it works! My skin and hair are healthier than ever. A super easy way, costs little money, and plastic-free!

CHINA and THE NETHERLANDS

Annemieke, Plastic Minimalism, China and the Netherlands.

Name: Annemieke
Website Name: PlasticMinimalism
Website URL: www.plasticminimalism.blogspot.nl

Several years ago, Annemieke came across a YouTube video of the talk I gave for an event called TEDxGreatPacificGarbagePatch, and the ideas stuck in her mind. During the years that followed, she increasingly read and heard about how plastic kills animals and ruins the planet. After seeing yet another photo series of bird stomachs filled with plastic, she decided to do something about it. So in 2013, she recalled that presentation she'd watch a few years before and realized that "what Plasticfree Beth could do, [she] could do too!"

Annemieke started her own Dutch blog, PlasticMinimalism, where she documents her steps towards a plastic-free life. And since she has lived in both the Netherlands and China, she's had to develop strategies for avoiding plastic in both countries. Here are a few of her tips. For more information and her in-depth plastic-free guide, please visit her website.

Some Plastic-Free Tips for China:

1. For drinking water, if you live in a compound, there is a big chance that there is a purified water vending machine. Buy a big glass jar at the supermarket (5 liter or so) and use this to get water. It's the cheapest option and plastic-free. If there is no such a vending machine in your area, get a water dispenser with refillable water jars that you order by telephone.

2. Organic veggies that are not wrapped in plastic are very hard to find. In Beijing, you might find some at the farmers market (北京有机农夫市集). At the Sanyuanli market (三源里市场) there is a saleswoman that sells organic produce. You'll find her in the back of the market.

3. When buying fruits and vegetables, make sure to bring your own reusable produce bags. In Chinese, tell the shop attendant "I don't want a plastic bag" = 我不要袋子 (wǒ bú yào dàizi) and offer your own bag at the same time. If you don't speak Chinese, bring along a small card with "I don't want a plastic bag" in Chinese. Because salespersons are so used to putting products in a plastic bag, it is normal that you have to repeat your message three times or so. If you buy just one item (i.e. one carrot, one apple), ask them to put the price tag straight on the product.

4. In general, markets are the best place to shop plastic-free. Often, especially in the case of fruit, the products are wrapped in plastic after arrival in the store. To get certain products plastic-free, ask the staff of the store to save some unpackaged fruit for you.

5. You can find nuts, beans, rice, wheat, dried fruit, and certain spices in bulk in the bigger grocery stores or at the local market. Bring your own bag or jars. If you buy meat, bring your own container. In Beijing, baking powder in cardboard boxes is available at expat stores (e.g. Jenny Lou's and April Gourmet). Vinegar is widely available in glass bottles.

In general, people in China are very cooperative when it comes to meeting your plastic-free ambitions. Everything is possible, just ask.

Some Plastic-Free Tips for the Netherlands:

1. Wherever you go, always carry an extra bag in your bag or car. If you forget, look for an empty cardboard box behind the cashiers to carry your products in.

2. Some plastic-free fruits and veggies are still sold at the mainstream supermarkets, for example, C1000 and AH. However, organic products in these stores are often wrapped in plastic. Go to organic stores to find a wider selection of plastic-free veggies and fruits. The chain store Ekoplaza is a great solution, as are markets. Gaia in The Hague (Aert van der Goesstraat 35-37) is also a good place to shop. Among the veggies that are especially hard to find plastic-free are spring onions and celery. AH sells spring onions plastic-free occasionally. Celery is virtually impossible to find plastic-free, unless it is local celery. Ask at the market or in the store when the season starts. Frozen veggies are sold in cardboard boxes at AH.

3. For fruits and veggies, bring your own reusable produce bags. You can buy Re-Sacks online or find them offline. Stores that sell them include GUUS at Rozengracht in Amsterdam and WAAR in The Hague.

4. Bring your own bags when buying bread, or buy (WASA) knäckebröd (crisp bread). Nuts, seeds, and dried fruit are available in bulk at some markets, for example, at the organic farmers market in The Hague (on Wednesdays). They are also available at Ekoplaza at the Weimarstraat in The Hague.

5. Beans, lentils, oats, rice, quinoa, and flour are also sometimes available in bulk at markets. Oats are available in a paper bag at AH (store brand). Check whether your local mill sells plastic-free flour.

6. Pasta, rice, and risotto are sold in cardboard boxes in many supermarkets. At some markets (for example, in Groningen on Saturdays), you will even find herbs, spices, and sun-dried tomatoes in bulk. In Utrecht, visit De Droom van Utrecht for herbs, spices, and tea in bulk. Bring your own jars.

7. Cheese can be bought plastic-free at cheese shops and markets. Bring your own container or bee's wrap. Other dairy products might be available in glass jars at the market or at an organic store. Or look up the closest dairy farmer for some really fresh, plastic-free milk. Yogurt you can make yourself. For meat and fish, bring your own container to the butchery or market. Most supermarkets sell honey and peanut butter in a jar with a plastic lid. Look for a local bee keeper or go to an organic store to find these plastic-free.

8. For household products like dish washing brushes and kitchen tools, visit Dille & Kamille. A large share of their collection is made from wood and stainless steel and does not have any packaging!

9. Baking soda is sold at Asian tokos (stores), for example at the chain store Amazing Oriental. Zaailing Natuurvoeding (Hoogracht 41) sells Ecover cleaning products in bulk. Bring your own jars or bottles. Detergent for washing clothes can be found in cardboard boxes. Soapnuts are nearly impossible to find plastic-free.

PORTUGAL

chelle Cassar, reusable mug ampion, Algarve, Portugal.

Name: Michelle Cassar

Website: None, but at the time of this writing, she and her friends are working on one.

I have known Michelle long distance via my blog since back when I called it "Fake Plastic Fish." Originally from England, Michelle has lived in Algarve in Southern Portugal since 2007, where she is a surfer and photographer. While out surfing, she'd often see bits of plastic floating by, and instead of ignoring the problem, would scoop them up and stuff them in her wetsuit to take back and recycle. And as a photographer, she'd spend summers shooting surf schools on different beaches, where bits of plastic would literally wash up at her feet. One evening, after a visit to clean up a local beach, she saw an article in ***The Surfer's Path*** magazine about ocean plastic pollution and a YouTube link to a video called "A Tale of Entanglement" about animals entangled in plastic. Michelle looked around at all the plastics in her camper and the next day started refusing and researching.

In 2011, I got a message from Michelle stating that she had refused over 10,000 single-use plastic items since she started, and she listed them out. It's helpful to tally up how much plastic we've used, but it's probably even more motivating to tally up what we haven't used.

I asked her about resources and challenges for living plastic-free in Portugal. She said things like spices, lentils, rice, and toilet paper are still difficult to find. Bread can be tricky because even though the local bread is delivered to the shop without plastic, it is then packaged in plastic by the shops. Her strategy is to ask if they have any bread in the back that hasn't been bagged yet. If they don't, she unties the wrapped bread and puts the bag back. Farmers markets are a great resource because not only is everything sold loose, but the food is great. Reducing her plastic means she has also started frequenting local shops more often, and it makes her feel good to support family businesses.

Recently, a small independent company called BeeCool Organics opened in the village. They offer personal care products like deodorant, lotions, sunscreen, lip balm, and even surf wax in plastic-free packaging and all made with beeswax from a local beekeeper. Products can be purchased in person or online through their website. Michelle told me when they realized that some of their ingredients came packaged in plastic, the owners looked into it and started ordering more, as larger quantities weren't packaged in plastic. They are focused on their supply chain, not just what you see on their shelves.

Michelle's other favorite plastic-free discover is her SteriPEN water filter, which she uses when travelling so she doesn't have to purchase bottled water. She says that during her trip to India, she never bought one bottle of water and never got sick. To prepare for her trip, she found a lot of the supplies she would need secondhand on eBay and requested no plastic packaging from the sellers.

One of Michelle's proudest accomplishments is helping Restaurante Praia Arrifana, a local beach restaurant/bar, to reduce its single-use disposable cup waste. The restaurant started using locally made mugs on their busy surfer evenings after Michelle suggested it to them. They implemented a deposit scheme to get the mugs back, or people can keep them as souvenirs. In the previous year, they started using glass bottles and giving out free water.

Her advice to people new to plastic reduction? Don't be embarrassed to ask for what you want. Ask with a smile. Be prepared to walk away if the answer is no, but usually the answer will be yes. Also, don't expect everyone you speak with to have the same "Aha!" moment you did. Do this for yourself and others will follow in their own way. If there aren't many people in your local vicinity who are as passionate as you, you'll find plenty of supporters (from around the world!) online.

UNITED KINGDOM

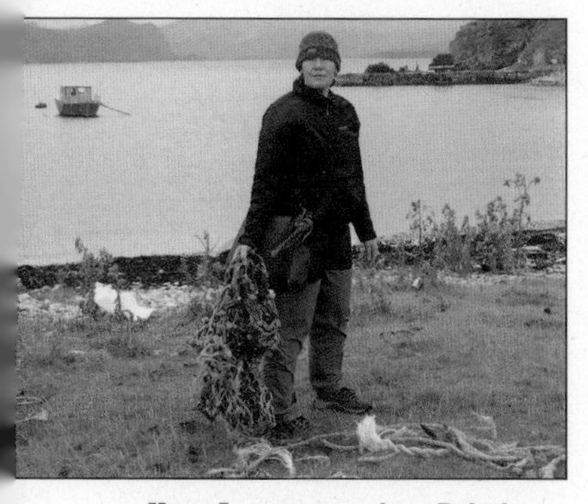

Kate Armstrong a.k.a. Polythene Pam, Huddersfield, UK.

Name: Kate Armstrong, a.k.a. Polythene Pam
Website Names: Plastic is Rubbish and Plastic-Free U.K.
Website URL: plasticisrubbish.com and plastic-free.co.uk

Kate lives in the UK with one husband and three chickens. They are not people who enjoy doing without life's pleasures. As Kate wrote me:

> We often shop at supermarkets, eat meat, drink alcohol, munch cheese, and scoff down cake. Giving up is not in our nature—we want to do everything—just without creating a huge pile of non-biodegradable, possibly carcinogenic, lethal rubbish that future generations will have to clean up.

They also like to travel a lot, and as I am writing this sentence, they are currently touring Europe in a van (the chickens are vacationing at a farm), and doing it plastic-free. So what made them decide to quit plastic? It started with a bag.

One day a plastic bag got tangled in the tree outside their house. Months later it was still there. The next year, when the leaves fell, there it was, looking all ragged and tatty and even more unpleasant. It was then Kate realized that plastic rubbish, unlike an apple core, doesn't biodegrade. It may seem obvious, but she had never considered it before, and, she thought, using a material that lasts forever to make disposable throw-away products had to be a misuse of plastic. In that moment, Kate realized that while she might not have been mindlessly scattering plastic litter herself, she was certainly misusing plastic. She too was a part of the problem.

So she started monitoring the amount of plastic trash she and her husband created. She saved all their plastic rubbish for a week, and by the end of those seven days, she was running out of space for it. Shocked, she decided to cut unnecessary plastics from their lives. It wasn't easy, especially when it came to buying food, but Kate's been at it for about eight years now, so she's developed a lot of strategies for people living in the UK and lists them on her website.

Here are a few of her tips:

1. For food, take your own plastic-free packaging and reusable bags and buy loose and unpackaged. To do this, you may have to step out of the supermarket and source a range of smaller, local shops. Of course, timetabling can be a real constraint when both parents work and supermarkets stay open all hours. However, taking the time to shop locally will not only cut your plastic footprint but also support the high street, and you might be surprised at how nice it actually is. Kate says that before going plastic free, she never really considered the role of the greengrocer in her life and now she sends him a Christmas card! One of the benefits of cutting packaging has been slow shopping, and she likes that.

   Coffeevolution in Huddersfield imports and roasts their own free trade coffee beans and will sell them loose and unpackaged if you ask them specially. Todmorden Market is a range of stalls selling loose tea, spices, olives, pies, and has the only unpackaged lettuce in Yorkshire. Wholefoods Market has branches in London, Glasgow, and Cheltenham. They sell dried foodstuffs loose.

2. Be flexible. Buy what loose veg, cheese, meat, and fish are available, and cook what you can with those ingredients. Don't decide what is for tea until you see what is loose on the shelves. You might have to learn to cook and eat less processed food. But is that really such a bad choice when ready-made foods not only come in plastic packaging but are often laden with extra salt, sugar, and unpronounceable flavorings and additives?

3. Develop a range of recipes suitable for plastic-free meals. Kate can always get leeks, courgettes (zucchini), and peppers plastic-free, so she's tracked down a whole load of recipes featuring those adaptable vegetables.

4. Grow some stuff. Salad is almost impossible to find plastic-free as are green beans. Lucky then that these things are really easy to grow.

5. Take it easy. Start by cutting one plastic-wrapped product a month, which will give you time to source an alternative and will make the task infinitely less daunting. It also gives others in the house time to adapt.

6. Enjoy finding alternatives. Picking your own strawberries is a great day out. For sure I can only eat lettuce in summer when I have grown it myself but when I do get it, it is a real treat. And home-grown does taste better. Homemade face cream is really pleasant to use, is infinitely cheaper, and makes a great gift.

7. Don't be overwhelmed. When you realize just how much plastic is out there you can easily break down and weep. But instead of counting the plastic you still use, look at what you have cut.

Kate's favorite plastic-free discoveries are milk delivered in glass bottles and plastic-free cocoa.

After many hours of researching plastic-free alternatives for herself, Kate wanted to make it easier for others to find resources. So she started her blog, Plastic is Rubbish, which contains an A–Z list of plastic-free products, and most recently a new website called Plastic-Free U.K., which is a directory of UK-based individuals and organizations tackling the plastic problem. Realizing that her point of view is not the only valid one and that she doesn't have all the answers, she wanted to create a site where people could discuss a variety of solutions and to show others that there is a market for plastic-free products and services and a growing concern about the problems of plastic abuse.

CANADA

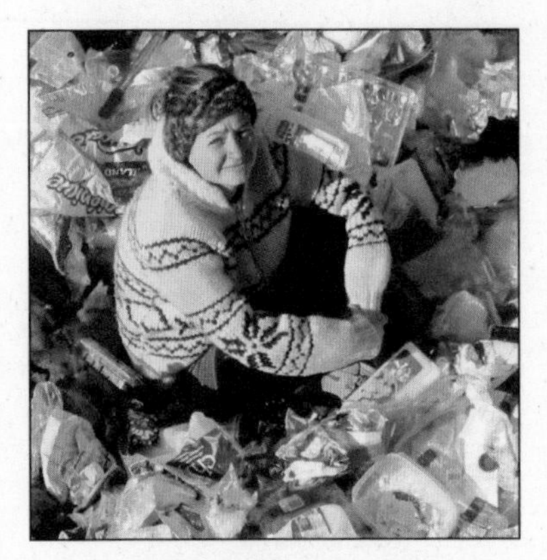

Taina Uitto, Plastic Manners, Vancouver, Canada

Name: Taina Uitto
Website Name: Plastic Manners
Website URL: PlasticManners.com and FromThe-WasteUpDoc.com

Five years ago, Taina saw a presentation by 5 Gyres that "rocked the way [she] saw the world." A few weeks later, she had an epiphany to quit plastic altogether and put an end to her "lazy consumer ways." She started a blog, Plastic Manners, to track her progress, and unbeknownst to many of us who met Taina back then, she also started filming much of her journey. Eventually, she and her brother turned their footage into a documentary film, ***From the Waste Up***, which is funny and provocative and shows what happens when several families in Vancouver, British Columbia, all try to give up plastic for a year. I highly recommend it!

When Taina first started, she really only had to worry about the plastic she and her then-boyfriend generated. Since then, she's had two kids (you can read about her method of reducing diaper waste through "elimination communication" on page 249), so she's had to figure out alternatives to baby-related plastic too.

Most of the resources Taina has found are already listed in this book since we're both, you know, on the same continent and all. But there are some local discoveries she loves. For example, in her city there is a store called the Soap Dispensary, which she describes as "basically a candy store of bulk/refill soaps and other cleaning things" like wood toothbrushes, for example. The store is expanding to kids' products like wooden

Lego. The regional district is also quite proactive. For example, they have a tap water app that shows you the nearby places that allow you to refill a water bottle. As for other stores, Taina sticks to smaller specialty stores (delis, butcher shops, etc.) where the experience is more personalized, the quality of the products is better, and plastic-free is possible.

She also says she gets a kick of out of plastic-free brushes and scrubbers. Without all the cheap colorful plastic junk, her kitchen doesn't look like a dollar store anymore.

But after five years of plastic reduction, Taina still has challenges. Milk is a big one. You can get it in glass, but the cap is always plastic. For health reasons, the paper and foil caps were outlawed, apparently because they weren't tamper proof (although she wonders how many tampering incidents there have ever actually been). Of course, even if the cap problem were solved, there would still be plastic in the milk (especially fatty milk) because of all the plastic tubing used in milk production. Yogurt is another challenge. After years of saying she will make her own, she still hasn't managed to do it. (Holy cow, woman, you're already doing so much!) She'd also like good quality dog kibble and for someone to come up with a plastic-free cap for bottled condiments.

Taina started her project as a personal challenge, which it still is, but couldn't stay quiet for long. And it wasn't just because she wanted to warn people of the horrifying effects of plastic pollution but because when she started to reap the benefits of living plastic-free, she wanted to share her knowledge with the world. In addition to writing her blog and making her movie, Taina accepts every presentation request and media opportunity.

INTERNATIONAL

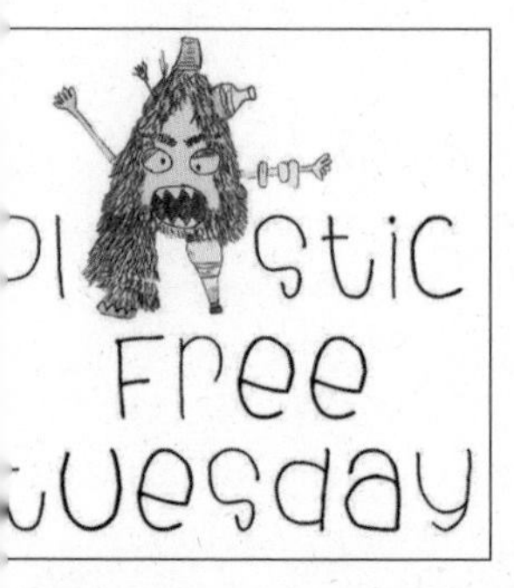

Website Name: Plastic-Free Tuesday
Website URL: plasticfreetuesday.com

I started this chapter with the international campaign called Plastic Free July. It seems appropriate to conclude it with another international initiative: Plastic-Free Tuesday.

Remember bi-national blogger Annemieke? Of course you do. You just read about her a few pages back. Well, after about a year of documenting her plastic reduction strategies on her blog, "Plastic Minimalism," she noticed that news about plastic pollution still kept appearing on television and in the newspapers. She wondered what she could do to get more people aware of the problem. And suddenly, the concept of Plastic-Free Tuesday came to her. Inspired by the success of Meatless Monday, she thought this would be an easy way for people to test what it is like to skip plastic.

Here's how it works: On Plastic-Free Tuesday, participants skip plastic to reduce their plastic footprint. They don't buy anything that is made of plastic or contains plastic. They also don't use anything made of plastic that they have to throw away after use. So, no bananas wrapped in plastic, no plastic bags, no take-away coffee in plastic cups, and so on.

Why just one day a week? Because, says Annemieke, adopting new habits takes time, especially if they are to last permanently. Pledging to give up plastic long-term may initially seem so daunting that many people would not even consider trying it. Plastic-Free Tuesday is a gateway drug to a plastic-free life (and a plastic-free world). Joining Plastic-Free Tuesday is a fun challenge, and more often than not, it is the start of a new life. Once you become aware of all the plastic that you consume and throw away in one single day, you will look at the world differently.

What starts as one plastic-free day a week may grow into permanent behavioral change. Once you have brought your own bags and jars to the grocery store on a Tuesday, it will start to feel strange to not do so on a Wednesday, Thursday, or Friday. Once you have had a Plastic-Free Tuesday, you will stop and think when standing in the produce section with a plastic-wrapped cucumber in front of you. And that reflection eventually makes the difference between mindless plastic shopping and refusing plastic.

Plastic-Free Tuesday was launched in early 2014 and has grown to an eight-member team and a handful of bloggers based across the globe. Currently, most of the campaign is online, but the team has plans for some offline activities in the future. You can join the campaign by not buying or throwing away any plastic on Tuesdays and sharing your Plastic-Free Tuesday successes on social media using #PlasticFreeTuesday.

This very short list of bloggers and activists is just the start. I'm acutely aware that it only includes participants from four of Earth's seven continents, and most of the world's countries are not represented. If you're out there working on reducing your personal plastic consumption and finding local resources where you live, please reach out and let me know, so we can connect and grow this movement together! None of us has all the answers individually. But together, we can be a mighty force for change.

Chapter 11: Nine Reasons Our Personal Changes Matter

During the 2008 presidential election campaign, Barack Obama was asked to name a personal green action he had taken. After the debate, he made an offhand remark about the question, which was published in *Newsweek* and made its way around the green blogosphere:

> So when Brian Williams is asking me about what's a personal thing that you've done [that's green], and I say, you know, "Well, I planted a bunch of trees." And he says, "I'm talking about personal." What I'm thinking in my head is, "Well, the truth is, Brian, we can't solve global warming because I f---ing changed light bulbs in my house. It's because of something collective."[117]

Since then, there have been a spate of articles arguing that while individual lifestyle changes and actions might make us feel good, they don't do much to solve the environmental problems we face. Our personal actions don't matter. Only policy changes can make a real difference. Obama and others who espouse this belief are right, and wrong. While simply changing light bulbs, carrying our containers to the grocery store, and tallying up our personal plastic waste are not enough to reverse the course of environmental degradation, our individual actions are not irrelevant. They matter. And here are nine reasons why.

Reason 1: When We Realize Our Direct Impact on the Rest of Life on the Planet, We Simply Cannot Continue to Do Harm. To Live with Integrity, We Have No Choice but to Change

Every day since June of 2007, I carry with me the mental image of an albatross chick that died with its belly full of the kinds of disposable plastic products that I used on a daily basis. Before that day, I lived in my own little bubble, insulated from feeling any but the most incidental connection to other creatures, or even people, that were not within my immediate sphere. One photograph changed all that. Suddenly I understood that the choices I make each day have impact beyond what I can imagine, and that we are all connected to each other, whether we are aware of those connections or not. Not in some philosophical, metaphysical way, but directly. The more we learn about the impact of our actions—how workers in Asia get sick from dealing with our plastic recycling, how workers in the United States get sick from manufacturing the chemicals for our plastic products, and how animals get sick from ingesting our plastic waste—we simply cannot continue to do harm. We change because it's the right thing to do. If that makes us feel good about ourselves, then hurray for an added bonus!

Reason 2: Our Actions Affect Our Own Health and the Health of Those We Love

While caring about animals was my entry point into the plastics issue, I've heard from other people—especially moms—that for them, the health of their children was their gateway. Reducing our exposure to the chemicals in plastic can have an immediate impact on our health. Since starting this project, I've learned to cook dried beans, grow and eat food from my own garden, and live on far fewer processed foods. I can't prove definitively that my health is better now than it was in June of 2007, nor can I prove that chemicals in plastics contributed to the reproductive health issues I faced back then, but I certainly feel more healthy now. And a recent study reported by the Breast Cancer

Fund found that when families that normally ate canned and packaged foods switched to eating unpackaged whole foods, the levels of BPA and the phthalate DEHP in their urine dropped by an average of 60 percent after only three days.[118] That's a huge payoff for a few personal changes. And, of course, eating whole foods has been found to have health benefits beyond simply reducing exposure to chemicals in plastics. Personally, reducing plastic was my gateway to living a healthier lifestyle.

Reason 3: We Can Vote with Our Wallets to Support Small, Ethical Businesses and Create a Greener Economy

Throughout this book, I've highlighted imaginative entrepreneurs creating products that are better for us and the planet than those offered by most mainstream corporations. These are people who had a good idea and did something about it. Companies like Life Without Plastic, Eco-Bags Products, ChicoBag, Green11, in.gredients, and Kippy's Ice Cream Shop are changing the way we think about buying products and carrying them home. The individual craftspeople of Etsy.com are finding ways to make money simply by doing what they already love to do. By hiring local people to repair our things when they break instead of chucking them out and replacing them, we promote an economy based on services rather than on extracting resources to produce more and more stuff. And by choosing not to spend money, simplifying our lives, and opting to barter and share rather than accumulate possessions, we can reduce the amount of time we have to spend making money to support lifestyles that might not have been making us so happy to begin with. Changing the way we spend our money helps create a more sustainable economy.

Reason 4: We Can Develop Our Own Ingenuity and Creativity and Learn How to Be More Self-Sufficient

To live without plastic either requires having a lot of money to pay people to create the plastic-free things I want or learning to make a lot of them myself. And since I'm not rich, I've had to learn a lot in the past five years. Learning to cook; to make my own mayonnaise, mustard, and chocolate syrup; to fix things when they break; to clean my house with everyday ingredients from the kitchen; to experiment and fail and try again and not give up have been frustrating at times but also exhilarating. More importantly, learning to figure out another way to obtain what I need rather than automatically accepting what is offered has broadened my experience of the world and what I'm capable of as a person. I fixed a freaking washing machine, and I wrote a book! What can't I do?

Reason 5: Personal Changes Lead Us to Examine Our Values and Evaluate What's Helpful to Our Physical and Spiritual Well-Being and What's Not

Choosing to live without buying new plastic has forced me to take a look at the things I relied on and question whether they were good for me in the first place. I tried—oh, I tried!—to find plastic-free frozen convenience foods. I was sad to say good-bye to Stouffer's macaroni and cheese and plastic bags of Pirate's Booty cheese snacks. But in the end, I had to recognize that those things were not only not good for my health, they weren't even making me happy. And it was more than just food. I would have loved to purchase an iPhone (maybe that's why I've already mentioned that particular item several times in this book), a Netflix player for my TV, and a few other techno gadgets, but in those moments of self-examination when I'm really being honest, I have to concede that getting more stuff never makes me happier—it just makes me want the next thing even more. It's a pleasure to refuse the treadmill of consumerism and focus on what really gives me pleasure: hanging out with my kitties, bantering with Michael, enjoying meals and movies with friends, spending time in my garden, learning and exploring and connecting with other people. The personal steps I took in June 2007 have led me to this point.

Reason 6: We Connect with Our Communities

Once we start making personal changes, most of us inevitably get to a point where we realize we can't change the world or even ourselves all alone. We run into problems and need advice. What's that weird smell in my dishwasher and how can I fix it? Why did my liquid soap turn out like green slime? Is there any brand of cotton swabs that doesn't come packaged in plastic? And if not, what can I use instead? Why are my squash plants shriveling up and dying? Are there any brands of tomato paste that come in a glass jar instead of a metal can? These are just a few of the questions I've posed to my online community through my blog and Facebook page, resources without which my plastic-free progress would have been much, much slower. There is a growing movement of people all trying to answer the same question: how can we reduce the amount of plastic in our lives? We don't all live in the same neighborhood, but through the wonder of modern technology (made possible by plastic!), we can find and connect with each other online. I've probably learned more from the people who read and comment on my blog than they have from me. Through the ongoing Show Your Plastic Trash Challenge, we share our successes and setbacks and help each other with ideas and suggestions.

And it's not just my *online* community that has grown. When I bring my own containers with me for take-out food or whip out my glass straw, I get to know the people who work in the local businesses around here. Instead of just ordering my sandwich and leaving, I begin a dialogue with them. They remember me from one visit to the next. That's not to say that you have to order your food in a stainless steel container to speak to the person serving your food, but those things can be conversation starters. The point is that the changes I've made in my personal life have led to more interpersonal connections than I'd probably made in my first forty-two years.

Reason 7: Hitting the Limits to Personal Changes Helps Us Recognize Where to Focus Our Energies in Asking Companies to Change, or to Start New Companies to Challenge the Status Quo

Until you actually start trying to live without plastic, you really have no idea what the stumbling blocks will be. What do you do when you want something that comes in plastic packaging, and you can't find it plastic-free anywhere else, you can't find it secondhand, you can't make it yourself, and you're not willing to live without it? That's when you realize you have to speak up and ask companies to change their practices. Living without plastic has helped me find my voice, a voice I never realized I had. It feels amazing to contact a company like LaundryTree about their packaging and have the owner not only give a positive response but actually make the changes you suggested. And it's indescribably empowering to ask a national brand like Clorox to make a change, face the company's opposition, rally the support of the community, work and persevere and see the changes you asked for actually carried out. The thing is, I never would have thought to lobby Brita to recycle its water filters if I hadn't been trying to get disposable plastic out of my personal life in the first place. My personal actions led directly to the greater community action that Obama was talking about in 2008. But the community action wouldn't have happened if I hadn't taken those first baby steps.

This is ***not*** to suggest that we all have to be as perfect as we can be before asking companies to change. Not at all. First of all, I don't believe in perfection. We are all just learning and stumbling along doing the best we can. In a conversation with Green Sangha founder Jonathan Gustin a while back, I mentioned that I thought it was hypocritical of us as individuals to blame corporations without looking at our own actions and the ways we support the companies polluting the environment. He replied, "It's okay to be a hypocrite."

Wow.

Jonathan went on to explain that while it's important for us to take personal action, we can't let that stop us from demanding change from the powers that be. Annie Leonard agrees. In her book, ***The Story of Stuff***, she writes:

> The individual actions we take to reduce our impact help us find the flaws in the system that need to be changed. I think of them as metal detectors leading us straight to what's wrong. Where the onus is on us individuals to do the right thing, these are the places in the system that need to be changed. . . . Why do I have to study GoodGuide for hours to figure out which shampoo, sunscreen, and lotions don't have carcinogens and reproductive toxics? System flaws! Instead, let's ban toxics in body care products so that everyone knows they are buying toxic free without investing hours of research.[119]

But it's those hours of research that reveal to us what's wrong with the system in the first place. And it's hitting those walls when we're trying to do the right thing that lets us know we have to go further.

The business people I interviewed in this book weren't satisfied with merely asking existing companies to change. Their personal actions—and if you'll notice, all of their stories began with individual changes—led them to decide to create alternatives to what already existed and to compete in the marketplace. The advice I heard in every interview was, "Just get started." If you have a great idea, don't wait for someone else to make it happen. Get out there and do it yourself!

Reason 8: We See the Flaws in the System

Sometimes lobbying companies doesn't work. Sometimes you can beg and plead and get thousands of people together to beg and plead, and companies won't budge. At that point, you need the force of law. You need systemic change on a massive scale. And boy oh boy, as I hope you've seen throughout this book, we need a lot of systemic changes. In my perfect world, we'd have:

- Fees and/or bans on single-use disposable bags, bottles, and plastic food-ware and packaging
- Commercial composting facilities and curbside pickup in all cities and mandatory composting and recycling for businesses
- Updated toxic chemical legislation that uses the Precautionary Principle as a basis for determining action on chemicals, requiring manufacturers to disclose the ingredients in their products and prove that they are truly safe *before* putting them on the market
- Extended Producer Responsibility legislation holding manufacturers accountable for the full life cycle of the products they produce
- Organic certification that takes the chemicals in plastic food packaging as seriously as the chemicals used to grow and process food

People like Renee Goddard and Andy Peri in Fairfax, California, Jean Hill in Concord, Massachusetts, the kids from Takoma Park, Maryland, and all the people taking a little time to get involved with campaigns through the Plastic Pollution Coalition, Green Sangha, Corporate Accountability International, Food & Water Watch, Sierra Club, Environment California, Surfrider Foundation, Environmental Working Group, Clean Water Action, Safer Chemicals Healthy Families, or the Campaign for Safe Cosmetics, give me hope that together we can make our voices loud enough to drown out those of corporate lobbyists and the big moneyed interests they represent.

In my perfect world, we'd see a new societal mindset. Our economies would be smaller, more local, and no longer based on the unsustainable idea of unlimited growth. We'd educate ourselves and our kids to watch commercials critically and to view marketing claims with a skeptical eye. We'd share more and work less. Or perhaps we'd work less at making money and more at doing the things we love.

Reason 9: By Letting Others See Our Personal Changes, We Set an Example of a Different Way to Be

When I see plastic trash on the sidewalk, I pick it up and hope other people see me doing it. No, I'm not trying to show off or appear virtuous. My hope is that by seeing someone pick up trash and dispose of it properly, others will get the idea that picking up trash is a normal thing to do: "Oh, people pick up trash on the sidewalk. It's okay to pick up trash. Maybe I'll do it too." It's why I've come to love the conversations I get into with people when I bring my own containers to stores and restaurants. "Oh," people say, "I didn't know you could do that." The more people see us with our reusable bags and water bottles, the more normal those things will become. Peer pressure works. We can use it for good or evil.

Setting an example through our personal actions is powerful. Eventually, we may want to take it further. For the past five years, I've challenged myself to speak up and speak out in various public forums: creating my blog, MyPlasticFreeLife.com, where I continue to write about plastic-free alternatives and challenges I encounter; giving presentations through Green Sangha's Rethinking Plastics program and eventually developing my own presentation, structured like the information in this chapter; visiting classrooms to talk to kids about plastic pollution; and even creating a crazy Plastic Sea Monster costume, which actually won the costume contest at San Francisco's annual Bay to Breakers race one year.

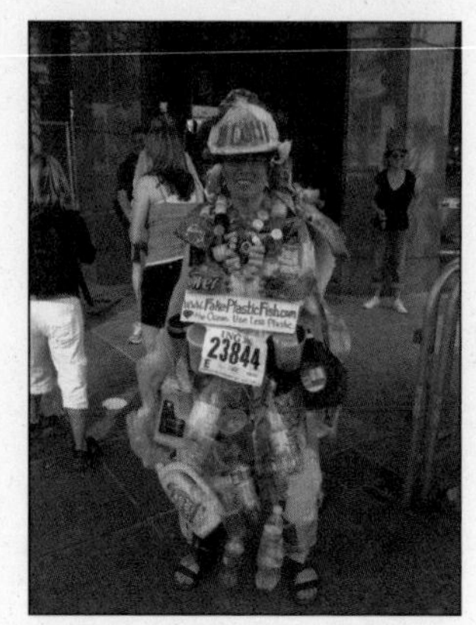

Plastic Sea Monster costume

The ones who really give me hope are kids setting an example for other kids as well as grownups: people like Milo Cress speaking up against disposable straws; nine-year old Anna Brooks from Takoma Park Young Activist Club, who will not take no for an answer when it comes to ditching disposable foodware; Jordan Howard, empowering other youth to speak up and educate each other; and the kids involved in school environmental clubs across the country, learning to take initiative, plan and run a campaign, and instigate tangible changes in their schools and communities. They are the ones who I'm counting on to help us shape a healthy future.

I started this chapter with a troubling quote from Barack Obama. I'll end with one that I take to heart:

> Change will not come if we wait for some other person or if we wait for some other time. ***We*** are the ones we've been waiting for. ***We*** are the change that we seek.[120]

Whoever you are, whatever your age, gender, or economic status, there is something for you to do in the fight against plastic pollution. There are so many ways to reach out and connect with the wider world. There are so many different ways to participate in this global movement. All talents and skills are needed.

Just pick one thing and get started.

Acknowledgments

Authors always say that their books could never have been written without the support of a lot of other people. In my case, that's absolutely true.

To my husband Michael Stoler, who not only read through my drafts and offered suggestions and advice, but also supported me through melt downs when I wasn't sure I could finish, took over my share of the housework, made sure our kitties had their homemade food each week, and kept us all safe from zombie invasion. He is truly my hero.

To my mom, who will never read this book; to my dad, who'd better read it; and to my brothers and sisters, who remind me that there's more to life than plastic!

To the people who believed in this book from the beginning: Jennifer Taggart, without whose example and encouragement I might never have written the first word; my sister Fran, who kicked me in the butt when I needed it; my agent Sarah Jane Freymann, whose vision has been in sync with mine from the start; and my editor at Skyhorse, Joseph Sverchek, whose idea it was to create this book without plastic—I never even had to ask.

To the trailblazers who paved the way for me: No Impact Man Colin Beavan and plastic-free blogger Envirowoman, who, as far as I know, was the first.

To the Women Eco-Warriors and to the community of MyPlastic FreeLife.com readers who contributed the plastic-free advice and tips in this book.

To the members of Green Sangha, particularly Stuart Moody, whose Rethinking Plastics training helped me find my voice; Jonathan Gustin, who helped me understand my purpose; and Elizabeth Little, whose steadfast support and friendship was invaluable.

To Jon Bernie, whose words helped me stay in the present moment through daily bouts of writing anxiety.

To my offline friends who have been there and supported me for years: I love you guys—even if you don't want to go shopping with me anymore. I sometimes feel uncomfortable shopping with me too.

To my workmates, in particular my boss Denise, who allowed me to take time off from my accounting job to write this book.

To those bringing attention to the plastic pollution problem: Captain Charles Moore of the Algalita Marine Research Foundation; Anna Cummins and Marcus Eriksen of 5 Gyres; Daniella Russo, Dianna Cohen, Manuel Maqueda, and Lisa Boyle, co-founders of the Plastic Pollution Coalition; and all the other plastic-free bloggers, activists and entrepreneurs who aren't waiting around for someone else to take action.

I thank you.

Notes

1 "Why Are Oil and Gas Fields a Concern?" NIH ToxTown website. http://toxtown.nlm.nih.gov/text_version/locations.php?id=150 (accessed 11/03/2014)

2 "Regulation of Hydraulic Fracturing Under the Safe Drinking Water Act," EPA website. http://water.epa.gov/type/groundwater/uic/class2/hydraulicfracturing/wells_hydroreg.cfm (accessed 11/2/2014)

3 Diamanti-Kandarakis, Evanthia, et al. "Endocrine-Disrupting Chemicals: An Endocrine Society Scientific Statement." ***Endocrine Reviews***. 2009. 30(4):293-342 www.endo-society.org/journals/scientificstatements/upload/edc_scientific_statement.pdf (accessed 12/1/2014)

4 Schmidt, C.W. "The Lowdown on Low-Dose Endocrine Disruptors." ***Environ Health Perspect***. 2001. 109:a420-a421. http://www.ncbi.nlm.nih.gov/pmc/articles/PMC1240453/pdf/ehp0109-a00420.pdf (accessed 12/1/2014)

5 The Endocrine Disruption Exchange. "Prenatal Origins of Endocrine Disruption." http://endocrinedisruption.org/prenatal-origins-of-endocrine-disruption/ (accessed 12/1/2014)

6 US EPA. "Toxicity and Exposure Assessment for Children's Health: Basic Information." www.epa.gov/teach/teachintro.html (accessed 12/1/2014)

7 US EPA. "Existing Chemicals: Bisphenol-A (BPA) Action Plan Summary." www.epa.gov/oppt/existingchemicals/pubs/actionplans/bpa.html (accessed 12/1/2014)
Safer Chemicals Healthy Families Resources: "Congressional Action Needed on a Chemical of High Concern: Bisphenol-A (BPA)." http://saferchemicals.org/get-the-facts/chemicals-of-concern/congressional-action-needed-on-a-chemical-of-high-concern-bisphenol-a-bpa/ (accessed 12/1/2014)

8 Colliver, Victoria, "Study: BPA, Methylparaben Block Breast Cancer Drugs." ***San Francisco Chronicle***. 13 Sep 2011. http://www.sfgate.com/news/article/Study-BPA-methylparaben-block-breast-cancer-2310172.php (accessed 12/1/2014)

9 US FDA. "Bisphenol A (BPA)." www.fda.gov/NewsEvents/PublicHealthFocus/ucm064437.htm (accessed 12/1/2014)

10 Yang, C.Z., et al. "Most Plastic Products Release Estrogenic Chemicals: A Potential Health Problem That Can Be Solved." ***Environ Health Perspect***. 2011. 119:989-996. http://ehp.niehs.nih.gov/1003220/ (accessed 12/1/2014)

Bittner, George D., et al. "Estrogenic chemicals often leach from BPA-free plastic products that are replacements for BPA-containing polycarbonate products." *Environmental Health 2014, 13:41.*
http://http://www.ehjournal.net/content/13/1/41 (accessed 12/1/2014)

11 Sax, Leonard. "Polyethylene Terephthalate May Yield Endocrine Disruptors." ***Environ Health Perspect***. 2010. 118:445-448.
http://ehp.niehs.nih.gov/0901253/ (accessed 12/1/2014)

12 Safer Chemicals Healthy Families Resources. "Congress Must Ensure Important Information about Chemical Use Is Not Hidden from People: Phthalates."
http://saferchemicals.org/get-the-facts/chemicals-of-concern/congress-must-ensure-important-information-about-chemical-use-is-not-hidden-from-people-phthalates/ (accessed 12/1/2014)

13 US Consumer Product Safety Commission. "Consumer Product Safety Improvement Act: Section 108."
http://www.cpsc.gov/Regulations-Laws--Standards/Statutes/The-Consumer-Product-Safety-Improvement-Act/Phthalates/FAQs-Bans-on-Phthalates-in-Childrens-Toys/ (accessed 12/1/2014)

14 Schade, Michael. "Toxic Toys 'R' Us: PVC Toxic Chemicals in Toys and Packaging." Nov 2010.
http://chej.org/wp-content/uploads/Documents/2010/ToxicToysRUs.pdf (accessed 12/1/2014)
Belliveau, Michael, and Stephen Lester. "PVC Bad News Comes in 3's: The Poison Plastic, Health Hazards and the Looming Waste Crisis." Dec 2004.
www.chej.org/wp-content/uploads/Documents/PVC/bad_news_comes_in_threes.pdf (accessed 12/1/2014)

15 Agency for Toxic Substances & Disease Registry. "Public Health Statement: Lead, CAS#: 7439-92-1." Aug 2007.
www.atsdr.cdc.gov/ToxProfiles/tp13-c1-b.pdf (accessed 12/1/2014)

16 Agency for Toxic Substances & Disease Registry. "Public Health Statement: Cadmium, CAS#: 7440-43-9." Sep 2008.
www.atsdr.cdc.gov/ToxProfiles/tp5-c1-b.pdf (accessed 12/1/2014)

17 US Consumer Safety Product Commission. "Consumer Product Safety Improvement Act: FAQs for Section 101."
http://www.cpsc.gov/en/Business--Manufacturing/Business-Education/Lead/FAQs-Lead-In-Paint-And-Other-Surface-Coatings/ (accessed 12/1/2014)

18 US Department of Health and Human Services. "Report on Carcinogens. 2011, 12 ed.
http://ntp.niehs.nih.gov/ntp/roc/twelfth/roc12.pdf (accessed 10/2/2014)

19 Royal Society of Chemistry. “Worrying levels of antimony found in popular fruit juices.” 01 Mar 2010.
www.rsc.org/AboutUs/News/PressReleases/2010/AntimonyFruitJuice.asp(accessed12/1/2014)

20 Helmenstine, Anne Marie. “Toxic Elements: Do You Know Which Elements Are Toxic?” About.com.
http://chemistry.about.com/od/toxicchemicals/a/toxicelements.htm (accessed 12/1/2014)

21 Choe, Suck-Young, et al. “Evaluation of estrogenicity of major heavy metals.” ***Sci Total Environ*** 2003. 312(1):15–21.
http://www.sciencedirect.com/science/article/pii/S0048969703001906(accessed12/1/2014)

22 Conversation with Robert Bateman, plastic bag manufacturer, Roplast Industries, Inc., 27 Aug 2010, who told me his bags contained “chicken fat.”
Further research turned up the following article: “Animal-Derived Agents in Disposable Systems: Growing Concern Over the Use of ADCs in Polymeric Materials.” ***Genetic Engineering & Biotechnology News***. 01 Aug 2005. Vol. 25, No. 14.
www.genengnews.com/gen-articles/animal-derived-agents-in-disposable-systems/1090/ (accessed 12/1/2014)

23 Helmut Kaiser Consultancy. “The Markets for Antimicrobial Additives / Biocides in Plastics and Textiles Worldwide 2010-2025: Development, Strategies, Markets, Companies, Trends, Nanotechnology.” Jun 2011.
www.hkc22.com/antimicrobials.html (accessed 10/1/2011)

24 Environmental Working Group. “EWG’s Guide to Triclosan.”
http://www.ewg.org/research/ewgs-guide-triclosan (accessed 12/1/2014)

25 Sanders, Robert. “Toxic Flame Retardants Found in Many Foam Baby Products.” ***UC Berkeley News Center***. 18 May 2011.
http://newscenter.berkeley.edu/2011/05/18/toxic-flame-retardants-found-in-many-foam-baby-products/ (accessed 12/1/2014)

26 Green Science Policy Institute.” Halogenated Flame Retardant Chemicals.”
http://greensciencepolicy.org/topics/health-environment/ (accessed 12/1/2014)

27 Nanrauskas, Vytenis, Arlene Blum, Rebecca Daley, and Linda Birnbaum. “Flame Retardants in Furniture Foam: Benefits and Risks.” Presented at 10th International Symposium on Fire Safety Science at the University of Maryland on 21 Jun 2011.
http://greensciencepolicy.org/wp-content/uploads/2013/12/Babrauskas-and-Blum-Paper.pdf (accessed 12/1/2014)

28 Environment News Service. “Chemicals in Fast Food Wrappers Show Up in Human Blood.”
www.ens-newswire.com/ens/nov2010/2010-11-08-01.html (accessed 12/1/2014)

29 U.S. EPA. “Long-Chain Perfluorinated Chemicals (PFCs) Action Plan Summary.”
www.epa.gov/oppt/existingchemicals/pubs/actionplans/pfcs.html (accessed 12/1/2014)

30 Science & Environmental Health Network. "Precautionary Principle." www.sehn.org/precaution.html (accessed 12/1/2014)

31 McDonald, G. Reid, et al. "Bioactive Contaminants Leach from Disposable Laboratory Plasticware," ***Science***. 07 Nov 2008. http://www.sciencemag.org/content/322/5903/917 (accessed 12/1/2014)

32 Martin, David S. "Toxic Towns: People of Mossville 'Are Like an Experiment.'" ***CNN Health***. 26 Feb 2010. www.cnn.com/2010/HEALTH/02/26/toxic.town.mossville.epa/index.html (accessed 12/1/2014)

33 Lerner, Steve. "Corpus Christi: Hillcrest Residents Exposed to Benzene in Neighborhood Next Door to Refinery Row." Collaborative on Health and the Environment, 24 Jul 2007. www.healthandenvironment.org/articles/homepage/1886 (accessed 12/1/2014)

34 Lee, G. Fred, PhD, and Anne Jones-Lee, PhD. "Flawed Technology of Subtitle D Landfilling of Municipal Solid Waste." G. Free Lee & Associates. Jul 2011. www.gfredlee.com/Landfills/SubtitleDFlawedTechnPap.pdf (accessed 12/1/2014)
Montague, Peter. "EPA Says All Landfills Leak, Even Those Using Best Available Liners." ***Rachel's Environment & Health News*** #37. 10 Aug 1987. www.rachel.org/files/rachel/Rachels_Environment_Health_News_1111.pdf (accessed 12/1/2014)

35 Popson, Colleen. "Museums: The Truth is in Our Trash." ***Archeology***, Volume 55, Number 1. Jan/Feb 2002. http://archive.archaeology.org/0201/reviews/trash.html (accessed 12/1/2014)

36 Environmental Integrity Project. "Waste-To-Energy: Dirtying Maryland's Air by Seeking a Quick Fix on Renewable Energy?" Oct 2011. www.environmentalintegrity.org/documents/FINALWTEINCINERATORREPORT-101111.pdf (accessed 12/1/2014)

37 GAIA. "Incinerators: Burning waste has many negative environmental, social, and health consequences." www.no-burn.org/section.php?id=84 (accessed 12/1/2014)

38 Allsopp, Michelle, Adam Walters, David Santillo, and Paul Johnston. "Plastic Debris in the World's Oceans." Greenpeace International. 09 November 2006. www.unep.org/regionalseas/marinelitter/publications/docs/plastic_ocean_report.pdf(accessed 12/1/2014)

39 Howshaw, Lindsey. "Afloat in the Ocean, Expanding Islands of Trash." ***The New York Times***. 9 Nov 2009. www.nytimes.com/2009/11/10/science/10patch.html (accessed 12/1/2014)

40 Eriksen, Mason, Wilson, et al. "Microplastic pollution in the surface waters of the Laurentian Great Lakes." Marine Pollution Bulletin, Volume 77, Issues 1–2. 15 December 2013.

41 The American Plastics Council (APC) and The Society of the Plastics Industry (SPI). "Operation Clean Sweep Pellet Handling Manual." http://www.opcleansweep.org/Manual (accessed 12/1/2014)

42 American Chemistry Council. "Plastic Packaging Resins." http://plastics.americanchemistry.com/Plastic-Resin-Codes-PDF (accessed 12/1/2014)

43 "Types of Plastics." PlasticsIndustry.com. www.plasticsindustry.com/types-plastics-p.asp (accessed 12/1/2014)

44 *Plastic Free Times.* "Plastic Pollution Kills Desert Animals, Too." http://www.theplasticfreetimes.com/news/10/07/27/plastic-pollution-kills-desert-animals-too (accessed 12/1/2014)

45 Byington, Cara. "The Bahamas: Crisis in the Checkout Line." The Nature Conservancy. http://www.nature.org/ourinitiatives/regions/caribbean/bahamas/howwework/whale-and-plastics.xml (accessed 12/1/2014)

46 Mrosovsky, N., G.D. Ryan, and M.C. James. "Leatherback Turtles: The Menace of Plastic." ***Marine Pollution Bulletin.*** 58: 287-289. 2009.

47 Telephone conversation with Wallace J. Nichols, 8/19/2011.

48 Boustead Consulting & Associates. "***Life Cycle Assessment for Three Types of Grocery Bags - Recyclable Plastic; Compostable, Biodegradable Plastic; and Recycled, Recyclable Paper.***" http://savetheplasticbag.com/UploadedFiles/2007%20Boustead%20report.pdf(accessed 12/1/2014)

49 O'Donnell, Jayne. "Tests Find High Levels of Lead in Reusable Bags." *USA Today.* 27 Jan 2011. http://usatoday30.usatoday.com/money/industries/retail/2011-01-23-reusable-bags_N.htm (accessed 12/1/2014)

50 Gerba, Charles P., David Williams, and Ryan G. Sinclair. "Assessment of the Potential for Cross Contamination of Food Products by Reusable Shopping Bags." 9 June 2010. http://www.llu.edu/assets/publichealth/documents/grocery-bags-bacteria.pdf(accessed 12/1/2014)

51 CalRecycle. "Annual Statewide Recycling Rates: Plastic Carryout Bags." www.calrecycle.ca.gov/Plastics/AtStore/AnnualRate/Default.htm (accessed 11/3/2014. There are no updated figures since 2009.)

52 There is some dispute as to whether plastic bags can actually be made from 100 percent recycled content. In an interview with Hilex Poly's Vice President of Sustainability & Environmental Policy, Mark Daniels, Stiv Wilson of 5 Gyres asserts that Daniels admitted to him that current technology only allows about 30 percent recycled content in plastic bags, and the rest must come from new plastic. Wilson, Stiv. "My Big Fat Problem with Recycling Plastic as a Solution to Plastic Pollution." 5Gyres.org. 29 Jun 2011.

www.5gyres.org/posts/2011/06/29/my_big_fat_problem_with_recycling_plastic_as_a_solution_to_plastic_pollution (accessed 12/1/2014)

53 Department of the Environment, Community and Local Government. www.environ.ie/en/Environment/Waste/PlasticBags/ (accessed 12/1/2014)
http://www.reuseit.com/facts-and-myths/the-plastax-irelands-plastic-bag-fee.htm (accessed 12/1/2014)

54 Riley, Charles. "No Paper, No Plastic. The Tax that Works Too Well." CNNMoney. http://money.cnn.com/2010/10/04/news/economy/DC_bag_tax/index.htm (accessed 12/1/2014)

55 Rogers, Rob. "Fairfax Folds on Plastic Bag Ban." ***Marin Independent Journal***. 30 Oct 2007. www.marinij.com/marin/ci_7318889 (accessed 12/1/2014)

56 Rogers, Rob. "Fairfax Bans Bags, Keeps Elected Clerk." ***Marin Independent Journal***. 04 Nov 2008. www.marinij.com/ci_10900602 (accessed 12/1/2014)

57 Container Recycling Institute. "Bottled Water." http://www.container-recycling.org/index.php/issues/bottled-water (accessed 11/3/2014)

58 US Government Accountability Office. "Bottled Water: FDA Safety and Consumer Protections Are Often Less Stringent Than Comparable EPA Protections for Tap Water." June 2009. www.gao.gov/new.items/d09610.pdf (accessed 12/1/2014)

59 Naidenko, Olga, PhD, et al. "Bottled Water Quality Investigation: 10 Major Brands, 38 Pollutants." Environmental Working Group. Oct 2008. http://www.ewg.org/research/bottled-water-quality-investigation (accessed 12/1/2014)

60 "Feng Shui/Bottled Water." ***Penn & Teller: Bullshit!*** Original air date March 7, 2003. (watched episode on YouTube on 09/16/2011)

61 Safeway. www.shop.safeway.com/superstore (accessed 09/16/2011)

62 Gas prices in the San Francisco Bay Area. www.sanfrangasprices.com (accessed 09/16/2011)

63 US Government Accountability Office. "Bottled Water: FDA Safety and Consumer Protections Are Often Less Stringent Than Comparable EPA Protections for Tap Water." June 2009. www.gao.gov/new.items/d09610.pdf (accessed 12/1/2014)

64 Food & Water Watch. "Fact Sheet: The Unbottled Truth About Bottled Water Jobs." June 2008. http://documents.foodandwaterwatch.org/doc/BottledWaterJobs.pdf (accessed 12/1/2014)

65 National Association for PET Container Resources. "Report on Postconsumer PET Container Recycling Activity in 2013." http://www.napcor.com/pdf/NAPCOR_2013RateReport-FINAL.pdf (accessed 11/3/2014)

66 Lauria, Tom. "The 7 Biggest Falsehoods About Bottled Water." ***Bottled Water Reporter***. Jun/Jul 2011. 38-39.
www.nxtbook.com/ygsreprints/IBWA/g20134ibwa_junjul11/#/40 (accessed 12/6/2014)

67 Alter, Lloyd. "SIGG Bottles Now BPA Free. But What Were They Before?" Treehugger. 20 Aug 2009.
http://www.treehugger.com/green-food/sigg-bottles-now-bpa-free-but-what-were-they-before.html (accessed 12/6/2014)

68 Fishman, Charles. "Message in a Bottle." Fast Company. 01 July 2007.
http://www.fastcompany.com/59971/message-bottle (accessed 12/6/2014)

69 Brentwood School water bottle project: Email correspondence with Will Bladt, Brentwood School middle division director. 08/13/2011 and 08/19/2011.

70 Hydration stations on college campuses:
Koch, Wendy and Kirsti Marohn. "BYOB and Refill at Water Stations." ***USA Today***. 15 Sep 2011. http://usatoday30.usatoday.com/news/education/story/2011-09-14/water-bottle-college/50403454/1 (accessed 12/6/2014)

71 Lohan, Tara. "Is Your City Going to Be Bottled Water-Free?" AlterNet. 24 June 2008. http://www.alternet.org/story/89148/is_your_city_going_to_be_bottled_water-free (accessed 12/6/2014)

72 Bundy on Tap. "Australia's First Bottled Water Free Town!" http://www.bundyontap.com.au (accessed 12/6/2014)

73 Thomas, Sarah. "Concord Town Meeting Defeats Bottled Water Ban." ***The Boston Globe***. 27 April 2011. http://www.boston.com/yourtown/news/concord/2011/04/concord_town_meeting_defeats_b.html (accessed 12/6/2014)

74 Telephone conversations with Jean Hill, on 9/17/2010 and 9/25/2011.

75 http://www.boston.com/yourtown/news/concord/2013/12/concord_rejects_repeal_of_plastic_water_bottle_ban.html (accessed 12/6/2014)

76 Tour of Davis Street Recycling Center with Rebecca Jewell on 10/8/2007, and follow-up email correspondences with Rebecca on 8/10/2011, 09/6/2011, 09/7/2011.

77 American Chemistry Council and The Association of Postconsumer Plastic Recyclers. "2013 United States National Post Consumer Plastics Bottle Recycling Report." 2014.
http://www.plasticsrecycling.org/images/pdf/resources/reports/Rate-Reports/National-Postconsumer-Plastics-Bottle-Recycling-Rate-Reports/2013_US_NATIONAL_POSTCONSUMER_PLASTIC_BOTTLE_RECYCLING_REPORT.pdf (accessed 11/7/2014)

78 Container Recycling Institute, "Bottled Up: (2000-2010) Beverage Container Recycling Stagnates, U.S. Container Recycling Rates & Trends, 2013." 2013. http://www.container-recycling.org/index.php/publications/2013-bottled-up-report (accessed 11/8/2014)

79 APR. "The Association of Postconsumer Plastic Recyclers Design for Recyclability Program." 2012.
http://www.plasticsrecycling.org/images/pdf/market_development/desig4recyclability/Full_APR_DFR.pdf (accessed 12/6/2014)

80 Williams, Holy. "Green Britain: Are You Poisoning China?" Sky News. 09 Jan 2007. http://news.sky.com/home/video/14160896 (accessed 07/Jan/2008)
Parry, Simon and Martin Smith, "The City That is Britain's Dustbin." ***Mail Online***. 02 Sep 2006. www.dailymail.co.uk/news/article-403348/The-city-Britains-dustbin.html (accessed 12/6/2014)

81 Jones, Richard. "Closed: The Chinese City Britain Uses as a Dustbin." ***Mail Online***. 27 Jan 2007.
www.dailymail.co.uk/news/article-431916/Closed-The-Chinese-city-Britain-uses-dustbin.html (accessed 12/6/2014)

82 MacEachern, Diane. "Michelle Bachman Needs to Go to China Before She Tries to Shut Down the EPA."***Huffington Post***. 22 Sep 2011. www.huffingtonpost.com/diane-maceachern/bachmann-epa_b_975899.html (accessed 12/6/2014)

83 Kirby, David. "Made in China: Our Toxic, Imported Air Pollution." ***Discover Magazine***. Apr 2011. http://discovermagazine.com/2011/apr/18-made-in-china-our-toxic-imported-air-pollution (accessed 12/6/2104)

84 Container Recycling Institute. "Bottled Up: (2000-2010) Beverage Container Recycling Stagnates, U.S. Container Recycling Rates & Trends, 2013."
http://www.container-recycling.org/index.php/publications/2013-bottled-up-report (accessed 11/8/2014)

85 U.S. EPA. "Municipal Solid Waste Generation, Recycling, and Disposal in the United States: Tables and Figures for 2012." Feb 2014.
www.epa.gov/solidwaste/nonhaz/municipal/pubs/2012_msw_dat_tbls.pdf (accessed 11/8/2014)

86 Franklin Associates. "Final Report: Life Cycle Inventory of 100% Postconsumer HDPE and PET Recycled Resin from Postconsumer Containers and Packaging." 7 Apr 2010.
www.napcor.com/pdf/FinalReport_LCI_Postconsumer_PETandHDPE.pdf (accessed 12/6/2014)

87 Kanellos, Michael. "Reaping Oil from Discarded Plastic." ***The New York Times***. 29 Sep 2011. http://green.blogs.nytimes.com/2011/09/29/an-oil-bonanza-in-discarded-plastic/(accessed 12/6/2014)

88 Lee, Karen. "Green Crafting: A Justifiable Means to an End?" Crafting a Green World. 2 Sep 2011.
www.craftingagreenworld.com/2011/09/02/green-crafting-a-justifiable-mean-to-an-end/ (accessed 12/6/2014)

89 U.S. EPA. "Municipal Solid Waste Generation, Recycling, and Disposal in the United States: Tables and Figures for 2012."
www.epa.gov/solidwaste/nonhaz/municipal/pubs/2012_msw_dat_tbls.pdf (accessed 12/6/2014)

90 American Chemistry Council and The Association of Postconsumer Plastic Recyclers. "2013 United States National Post Consumer Plastics Bottle Recycling Report." 2014.

http://www.plasticsrecycling.org/images/pdf/resources/reports/Rate-Reports/National Postconsumer-Plastics-Bottle-Recycling-Rate-Reports/2013_US_NATIONAL_POSTCONSUMER_ PLASTIC_BOTTLE_RECYCLING_REPORT.pdf (accessed 11/7/2014)

91 Port of Oakland. "Recycling/Waste Reduction."
http://oaklandairport.com/noise/environmental_recycle.shtml (accessed 12/6/2014)

92 US EPA. "Municipal Solid Waste Generation, Recycling, and Disposal in the United States: Tables and Figures for 2012."
www.epa.gov/solidwaste/nonhaz/municipal/pubs/2012_msw_dat_tbls.pdf (accessed 12/6/2014)

93 Caltrans. "California Department of Transportation District 7 Litter Management Pilot Study." Caltrans Document No. CT-SW-RT-00-013, 26 June 2000.
www.dot.ca.gov/hq/env/stormwater/pdf/CTSW-RT-00-013.pdf (accessed 12/6/2014)

94 Californians Against Waste. "Polystyrene: Local Ordinances."
www.cawrecycles.org/issues/plastic_campaign/polystyrene/local (accessed 11/16/2014)

95 Portland Schools Durable Trays Pilot Program:
www,pps.k12.or.us/departments/facilities/3247.htm (accessed 11/16/2014)

96 NY City Trayless Tuesdays Program.
www.sosnyc.org/Trayless_Tuesday.html (accessed 12/6/2014)

97 Piney Branch Elementary School pushing for washable trays:
Telephone conversation with Young Activist Club leader Brenda Platt 06/10/2011.
Telephone conversation with Young Activist Club member Anna Brooks 06/25/2011.
Email update from Young Activist Club leader Brenda Platt 11/16/2014.

98 Plastic coatings themselves can break down into tiny micro-plastic particles:
Eco-Cycle. "Frequently Asked Questions About Microplastics in Compost."
www.eco-cycle.org/microplasticsincompost/faqs.cfm (accessed 10/3/2011)
http://www.ecocycle.org/specialreports/microplasticsincompost (accessed 12/6/2014)

99 The standards for testing biodegradability are as follows:

Biodegradability under composting conditions

- Specification Standards ASTM D6400, D6868 (coatings)
- Specification Standards EN 13432 (European Norm)
- Specification Standards ISO 17088 (International Standard)

Biodegradability under marine conditions

- Specification Standard D 7021

Biodegradability Test Methods—ASTM Standards

- Soil D5988
- Anaerobic digestors D 5511, ISO 15985
- Biogas energy plant

• Accelerated landfill D 5526

• Guide to testing plastics that degrade in the environment by a combination of oxidation and biodegradation ASTM D 6954

Source: Narayan, Ramani. "Understanding ASTM Standards in the Bioplastics Area: Science Behind Biodegradable-Compostable Plastics." Michigan State University. 27 Jul 2011.

100 Per email exchange with Hain-Celestial Customer Service. 5 July 2013

101 Sutton, Rebecca, PhD. "Teen Girls' Body Burden of Hormone-Altering Cosmetics Chemicals." Environmental Working Group. Sep 2008. http://www.ewg.org/research/teen-girls-body-burden-hormone-altering-cosmetics-chemicals (accessed 12/6/2014)

102 Houlihan, Jane. "Why This Matters—Cosmetics and Your Health." Environmental Working Group. http://www.ewg.org/skindeep/2011/04/12/why-this-matters/ (accessed 12/6/2014)

103 U.S. EPA. "Municipal Solid Waste Generation, Recycling, and Disposal in the United States: Tables and Figures for 2012." http://www.epa.gov/osw/nonhaz/municipal/pubs/2012_msw_dat_tbls.pdf(accessed12/06/2014)

104 Aumônier, Simon, Michael Collins, and Peter Garrett. "Science Report—An Updated Life-cycle Assessment Study for Disposable and Reusable Nappies." Oct 2008. https://www.gov.uk/government/uploads/system/uploads/attachment_data/file/291130/scho0808boir-e-e.pdf (accessed 12/6/2014)

105 Gehring, M., Tennhardt, L., Vogel, D., Weltin, D., Bilitewski, B. "Bisphenol A Contamination of Wastepaper, Cellulose and Recycled Paper Products." Waste Management and the Environment II. WIT Transactions on Ecology and the Environment, vol. 78, Southampton, Boston: WIT Press, 294—300. http://rcswww.urz.tu-dresden.de/~gehring/deutsch/dt/vortr/040929ge.pdf (accessed 12/6/2014)

106 Syufy, Franny."Cat Litter—To Scoop or Not to Scoop: The Clumping Clay Controversy." About.com. http://cats.about.com/cs/litterbox/a/clumpingclay.htm (accessed 12/6/2014)

107 The safety of flushing cat litter from indoor cats who have tested negative for toxoplasma gondii was confirmed via email by an NRDC representative on 01/28/2008.

108 Belliveau, Michael and Stephen Lester. "PVC Bad News Comes in 3's: The Poison Plastic, Health Hazards and the Looming Waste Crisis." Center for Health, Environment, and Justice and Environmental Health Strategy Center. Dec 2004. www.chej.org/wp-content/uploads/Documents/PVC/bad_news_comes_in_threes.pdf (accessed 12/6/2014)

109 Rust, Susanne and Meg Kissinger. "BPA Leaches from 'Safe' Products." *Journal Sentinal*. 15 Nov 2008.

www.jsonline.com/watchdog/watchdogreports/34532034.html (accessed 12/6/2014)

110 U.S. EPA. "Research Investigates Human Health Effects of Nanomaterials." www.epa.gov/nanoscience/quickfinder/hh_effects.htm (accessed 12/6/2014)

111 Syufy, Franny. "Feline Acne, AKA 'Kitty Acne' or 'Chin Acne'." http://cats.about.com/cs/healthissues/a/acne.htm (accessed 12/6/2014)

112 Cliver, Dean O., PhD. "Plastic and Wooden Cutting Boards." http://faculty.vetmed.ucdavis.edu/faculty/docliver/Research/cuttingboard.htm(accessed 12/6/2014)

113 "Silicone Tally: How Hazardous Is the New Post-Teflon Rubberized Cookware?" ***Scientific American***. 5 May 2010. http://www.scientificamerican.com/article/earth-talk-silicone-tally/ (accessed 12/6/2014)

114 Jackson, Nicholas. "Here's Why People Will Buy Apple's New White iPhone 4." ***The Atlantic***. 26 Apr 2011. www.theatlantic.com/technology/archive/2011/04/heres-why-people-will-buy-apples-new-white-iphone-4/237890/ (accessed 12/6/2014)

115 Raloff, Janet. "Synthetic Lint Ends Up In Oceans." ***Science News***. 14 Sep 2011. https://www.sciencenews.org/article/synthetic-lint-ends-oceans (accessed 12/6/2014)

116 prAna Blog. "prAna is Taking Packaging Back to Basics." 20 Feb 2011. www.prana.com/blog/2011/02/20/prana-is-taking-packaging-back-to-basics/ (accessed 02/22/2012)

117 "How He Did It." ***Newsweek***. 17 Nov 2008. p.38. http://www.newsweek.com/barack-obama-how-he-did-it-85083 (accessed 12/6/2014)

118 Rudel R.A., et al. "Food Packaging and Bisphenol A and Bis(2-Ethyhexyl) Phthalate Exposure: Findings from a Dietary Intervention." ***Environ Health Perspect.*** 2011. 119:914-920. http://ehp.niehs.nih.gov/1003170/ (accessed 12/6/2014); Breast Cancer Fund. "BPA in Food Packaging Study." http://www.breastcancerfund.org/clear-science/research-methods/food-packaging-study/ (accessed 12/6/2014)

119 Leonard, Annie. ***The Story of Stuff***. New York: Free Press, 2010. p.264.

120 Barack Obama's Feb 5 Speech, ***The New York Times***. 05 Feb 2008. www.nytimes.com/2008/02/05/us/politics/05text-obama.html (accessed 12/6/2014)

Index